AF598390

COMBINATORIAL STRATEGIES in BIOLOGY and CHEMISTRY

Combinatorial Strategies in Biology and Chemistry

Annette Beck-Sickinger
University of Leipzig, Germany

Peter Weber
University of Leipzig, Germany

Translated by
Michael Soderman and **Allan Wier**

JOHN WILEY & SONS, LTD

Baffins Lane, Chichester,
West Sussex, PO19 1UD, England

National 01243 779777
International (+44) 1243 779777

e-mail (for orders and customer service enquiries): cs- books@wiley.co.uk

Visit our Home Page on http://www.wiley.co.uk or http://www.wiley.com

Other Wiley Editorial Offices

John Wiley & Sons, Inc., 605 Third Avenue,
New York, NY 10158–0012, USA

WILEY-VCH Verlag GmbH
Pappelallee 3, D-69469 Weinheim, Germany

John Wiley & Sons Australia, Ltd
33 Park Road, Milton, Queensland 4064, Australia

John Wiley & Sons (Canada) Ltd, 22 Worcester Road,
Rexdale, Ontario, M9W 1L1, Canada

John Wiley & Sons (Asia) Pte Ltd, 2 Clementi Loop #02–01,
Jin Xing Distripark, Singapore 129809

Library in Congress Cataloging-in-Publication Data

Beck-Sickinger, Annette.
[Kombinatorische Methoden in Chemie und Biologie. English]
Combinatorial strategies in biology and chemistry / Annette Beck-Sickinger, Peter Weber; translated by Allan Wier.
p. cm.
Includes bibliographical references and index.
ISBN 0-471-49726-6 (acid-free paper) - ISBN 0-471-49727-4 (pbk.: acid-free paper)
1. Combinatorial chemistry. I. Weber, Peter. II. Title.

RS419 .B4313 2001
615.19—dc21

2001026499

British Library Cataloguing in Publication Data

A catalogue record for this book is available from the British Library

ISBN 0-471-49726–6 (Cloth)
ISBN 0-471-49727–4 (Paper)

Typeset in 10/12pt Times by Kolam Information Services Pvt. Ltd, Pondicherry, India.
Printed and bound in Great Britain by Biddles Ltd, Guildford and King's Lynn.
This book is printed on acid-free paper responsibly manufactured from sustainable forestry, in which at least two trees are planted for each one used for paper production.

Contents

Preface

Combinatorial methods are revolutionizing the discovery of substances with outstanding properties. While the traditional principle focuses on the synthesis of one individual substance at a time and the subsequent screening of its properties, the new methodologies aim at the identification of one or a few very promising candidates through the parallel or mixture-based synthesis and screening of a multitude of compounds. Even though this approach is still new, it strongly influenced the synthetic strategies in all kinds of laboratories. This is particularly true for industrial research, but also affects universities as well as scientific research institutes.

When new methods arise, it typically takes far too long for these techniques to get included in publications, especially textbooks. In the present case this means that there is still a lack of books that offer a basic and comprehensive introduction to combinatorial methods in chemistry and biology for undergraduate and graduate students as well as natural scientists in neighboring fields. This was motivation enough for us to tackle the preparation of this textbook.

As a starting point, we used lecture manuscripts as well as introductory presentations about the topic in which we tried to include examples of current scientific research. In doing so, we noticed several times in a variety of publications that the same terms were related to completely different things. Thus we decided to document the diverse directions of development starting from the very beginning. In this context it is impossible to provide a complete coverage of the topic as a matter of course. Every week, a multitude of articles are published in almost all leading scientific journals. Furthermore, today there exist a number of journals that focus exclusively on results that were produced by combinatorial methods or on the development of combinatorial techniques. We therefore tried to clarify the principles, to sort the methods and to structure the whole field in order to provide the reader with a solid base for the understanding and interpretation of current and future results.

Since combinatorial methods have particularly gained importance for the pharmaceutical and chemical industries we deliberately sought examples of applications in these areas and managed to obtain them thanks to the support of a number of colleagues. Without this legwork the illustration and practical relevance would not have been as authentic especially in the chapter about automation. We therefore wish to thank accelab (Kusterdingen, Germany), Jerini (Berlin, Germany), MultiSynTech (Witten, Germany), Rapp Polymere

(Tübingen, Germany), and Tecan (Männedorf, Switzerland) for providing us with information and figures about the latest developments in the field of automation of combinatorial syntheses. Furthermore, we are grateful to Drs Bovermann, Falchetto, and Yang (Novartis, Basel/Summit, Switzerland/USA) for their practical examples of applications in the field of industrial syntheses and analyses based on combinatorial methods.

We also want to thank our colleagues who supported, influenced and informed us by discussing various aspects of this field and providing helpful information about the topic. I (A. G. B.-S.) am particularly grateful to my scientific teachers Dr R. A. Houghten (Torrey Pines Institute for Molecular Studies, San Diego, USA) [I did my first parallel syntheses in his laboratory at that time at The Scripps Research Institute (La Jolla, USA)], Professor Dr G. Jung (Eberhard-Karls-University, Tübingen, Germany), and Professor Dr G. Folkers (ETH, Zürich, Switzerland) for their support and the freedom they gave me for my personal development, which is a prerequisite for young scientists entering the fascinating field of scientific research.

Annette Beck-Sickinger
Peter Weber
Leipzig, May 2001

List of Abbreviations

AA	amino acid
ACE	(1) affinity capillary electrophoresis or (2) angiotensin-converting enzyme
Acm	acetamidomethyl
ADCC	4-acetyl-3,5-dioxo-1-methylcyclohexane carboxylic acid
All	allyl
Aloc	allyloxycarbonyl
AmpR	ampicillin resistance gene
ATP	adenosine triphosphate
ATR	attenuated total reflection
ATZ	anilinothiazolinone
Boc	*tert*-butyloxycarbonyl
BOP	benzotriazole-1-yl-oxy-tris(dimethylamino)-phosphonium hexafluorophosphate
CE	capillary electrophoresis
CLND	chemiluminescent nitrogen detector
CPC	controlled pore ceramics
CPG	controlled pore glass
DCM	dichloromethane
DCR	divide, couple, and recombine
Dde	1-(4,4-dimethyl-2,6-dioxocyclohexylidene)ethyl
ddNTP	2′,3′-dideoxynucleotide
Ddz	α, α-dimethyl-3,5-dimethoxybenzyloxycarbonyl
DIC	1,3-diisopropylcarbodiimide
DIEA	*N,N*-diisopropylethylamine
DIN A4	German industrial standard A4
DMAP	4-dimethylaminopyridine
DMF	*N,N*-dimethylformamide
DMSO	dimethylsulfoxide
DMT	dimethoxytrityl
DNA	deoxyribonucleic acid
dNTP	2′-deoxynucleotide
DsbA	disulfide bond-forming enzyme
DTT	1,4-dithio-DL-threitol
DVB	divinylbenzene
EGTA	ethylenebis(oxyethyleneitrilo)tetraacetic acid

ELISA	enzyme-linked immunosorbent assay
ELSD	evaporative light-scattering detector
ESI	electrospray ionization
FACS	fluorescence-activated cell sorter
Fmoc	9-fluorenylmethoxycarbonyl
FT	Fourier transform
GC	gas chromatography
GFP	green fluorescent protein
H-chain	heavy chain
HATU	*O*-(7-azabenzotriazol-1-yl)-1,1,3,3-tetramethyluronium hexafluorophosphate
HBTU	*O*-(benzotriazol-1-yl)-1,1,3,3-tetramethyluronium hexafluorophosphate
HF	hydrogen fluoride
HIPE	high internal phase emulsion
HMBA	4-hydroxymethylbenzoic acid
HMP	4-hydroxymethylphenol
HOBt	1-hydroxybenzotriazole
HPLC	high-performance liquid chromatography
HSQC	heteronuclear single-quantum coherence
HTS	high-throughput screening
IC_{50}	half-maximal inhibition constant
ICR	ion cyclotron resonance
IgG	immunoglobulin G
IR	infrared spectroscopy
kDa	kilodalton
L-chain	light chain
LC	liquid chromatography
LOSC	laser optical synthesis chip
m/z	mass/charge
MALDI	matrix-assisted laser desorption/ionization
MAS	magic angle spinning
MBHA	4-methylbenzhydrylamine
MHC	major histocompatibility complex
Moz	4-methoxybenzyloxycarbonyl
MPS	multiple peptide syntheses
mRNA	messenger ribonucleic acid
MS	mass spectrometry
NHS	*N*-hydroxysuccinimide
NMP	*N*-methylpyrrolidone
NMR	nuclear magnetic resonance
NPY	neuropeptide Y
Nvoc	6-nitroveratryloxycarbonyl

ODmab	4-{*N*-[1-(4,4-dimethyl-2,6-dioxocyclohexylidene)-3-methylbutyl]amino}benzyloxy
PAL	peptide amide linker
PAM	phenylacetamidomethyl
PAMAM	polyamidoamine
PCR	polymerase chain reaction
PE	polyethylene
PEG	polyethylene glycol
PEGA	polyethylene glycol dimethyl acrylamide
PFG	pulsed field gradient
PG	protecting group
PITC	phenylisothiocyanate
Piv	pivaloyl
PMB	*p*-methoxybenzyl
POEPOP	polyethylene glycol derivatized with epichlorohydrin
POEPS	polyethylene glycol cross-linked oligostyrene
Pon	acyl-2-[(oxymethyl)phenylacetoxy]-propionyl
Pop	[acyl-4-(4-oxymethyl)phenylacetoxy-methyl]-3-nitrobenzoic acid
PPTS	pyridinium *p*-toluenesulfonate
PS	polystyrene
PSD	post-source decay
PTC	phenylthiocarbamyl
PTH	phenylthiohydantoin
PyBOP	benzotriazole-1-yl-oxy-tris-pyrrolidino-phosphonium hexafluorophosphate
REM	regenerated (...) Michael addition
RNA	ribonucleic acid
RNAse	ribonuclease
RT	room temperature
SAR	structure – activity relationship
SASRIN®	super acid sensitive resin
SCF	supercritical fluid
SELEX	systematic evolution of ligands by exponential enrichment
SIMS	secondary ion mass spectrometry
SMART™	single or multiple addressable radio-frequency tag
SMPS	simultaneous multiple peptide syntheses
SPOCC	polyethylene glycol derivatized with oxetane
TBTU	*O*-(benzotriazol-1-yl)-1,1,3,3-tetramethyluronium tetrafluoroborate
*t*Bu	*tert*-butyl
TFA	trifluoroacetic acid
TFMSA	trifluoromethyl sulfonic acid
THF	tetrahydrofuran

THP	tetrahydropyranyl
TOF	time-of-flight
tRNA	transfer ribonucleic acid
trNOE	transferred nuclear Overhauser effect
UV	ultraviolet

1 Introduction

1.1 WHAT DOES COMBINATORICS MEAN?

Over the past few years, a single catch phrase has caused a stir with a series of chemical as well as biological questions: 'Combinatorics'. What do we mean by that? In mathematics, combinatorics is the theory of the number of different possible arrangements of given things or elements. According to the dictionary, combination means a linking of several simple forms. Terms mentioned in the same breath with combinatorial syntheses are compound libraries, mixture syntheses, combinatorial chemistry and diversomers.

So what is behind all these terms, which found their way into the language usage of chemists and biologists over the past few years? Let's consider an example: A company has discovered an interesting protein, and one suitable ligand would possibly be a new active compound for the therapy of a disease. How can the company proceed? It can characterize the protein as precisely as possible, manufacture it in large amounts through gene-technological methods, crystallize it and determine the three-dimensional structure using X-ray analysis. Proceeding from this structure it can now conceive possible ligands, then synthesize them and investigate their effect in biological assays. Through many synthesis and test cycles, the ligands can be optimized in this way. One successful example of this procedure is the development of a class of hypotensive medicines, the so-called ACE inhibitors (inhibitors of the angiotensin-converting enzyme).

A second possible procedure for the company would be searching for an endogenous ligand, in order to begin the optimization starting from this ligand. For this purpose, synthetic variants of the ligand are manufactured with the aim of giving information about the relevant interactions with the protein. In general, there follow many synthesis and test cycles, until a so-called lead structure is found, whose optimization then results in the ligand with the desired properties.

For some time now, there has been a third path being taken in the search for active compounds: high-throughput screening. In the identification of a new target molecule, for example a receptor, all compounds that have ever been manufactured in the company undergo specific biological test procedures – in the hope of finding a molecule that happens to interact with the target molecule. As an alternative or in addition, natural compounds, such as plant extracts, secondary metabolites from bacteria, fungi or marine organisms are examined for the desired properties.

To boost the variety of compounds, now there is a fourth path that can be taken: combinatorial synthesis, that is, the synthesis of a variety of molecules by the combination of different building blocks in different ways. These compounds can then undergo parallel or sequential examination, to find a lead structure or optimize an existing structure.

The idea that the test system itself searches for its active molecule from a series or a mixture of compounds is fascinating and promising, but not new. Every day, our bodies proceed millions of times in accordance with precisely this principle: receptors on cell surfaces recognize their ligands from thousands of compounds. This procedure is even more important in our immune system: antibodies detect protein fragments of bacterial or viral surface structures, out of millions of protein fragments, major histocompatibility complex (MHC) molecules bind these with very special anchor sequences. Thus, the application of combinatorial methods is not limited to compound research, even though this still does represent the greatest field of application. For example, in inorganic chemistry, catalysts are optimized by combinatorial methods, and in biochemistry as well as in molecular biology, combinatorial methods are employed to answer diverse questions.

1.2 THE STRUCTURAL ELEMENTS OF COMBINATORICS

If we take another look at the definitions, they are always talking about simple elements or building blocks that can be combined in various ways. Nature has known this principle for a long time: Four different nucleotides polymerize through formation of phosphoric acid ester bridges to linear ribonucleic acid (RNA) or deoxyribonucleic acid (DNA) molecules (Figure 1.1), and 20 different amino acids polycondense through dehydration under formation of amide bonds to linear peptides and proteins (Figure 1.2). A nucleic acid of 300 base pairs has 4^{300} different sequence possibilities. Since each of the three successive base pairs defines one of the 20 natural amino acids, 300 base pairs can be translated into 20^{100} different proteins – an incredible variety!

In addition, nature combines whole gene sections through rearrangement on different levels. The formation of the huge variety of antibodies is one example (Figure 1.3): The antibody binding sites are composed of the variable regions of the heavy chain (H-chain) and the light chain (L-chain). The variable region of the κ-L-chain (κ is the largest family of the L-chains) is coded by approximately 250 V-gene segments and four J-gene segments, whereby a V-segment can be combined with a J-segment in at least three different ways. This results in at least $250 \times 4 \times 3 = 3000$ combination possibilities for the encoding of the κ-L-chain. The variable region of the H-chains is coded by five instead of four J-gene segments and additionally by about 15 D-gene segments, so that there are at least $250 \times 5 \times 15 \times 3 = 56\,250$ combination possibilities for the encoding of the H-chain. The association of H-chains and L-chains then theoretically results

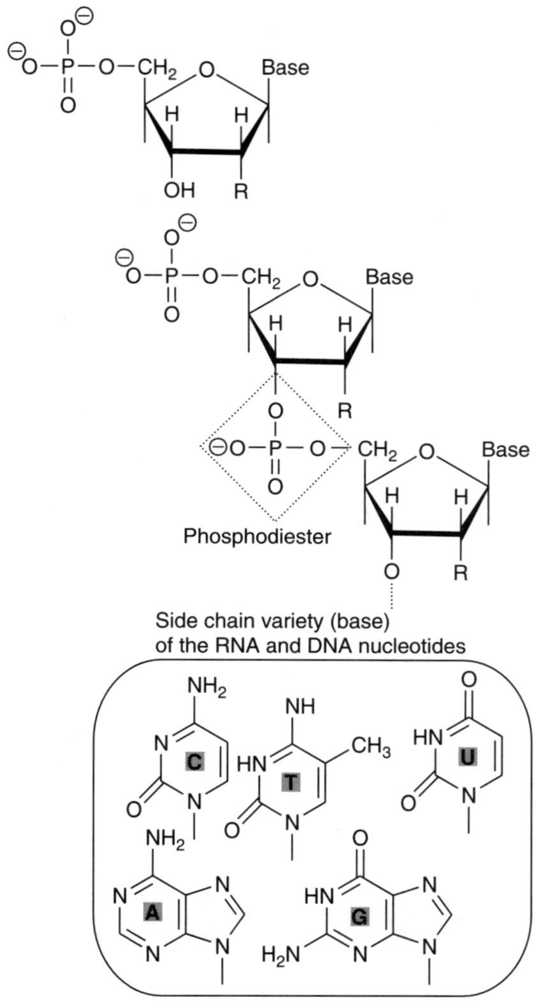

Figure 1.1 (Deoxy-) nucleotides are the building blocks of nucleic acids. For R = OH one speaks of ribonucleic acids (RNA), for R = H one speaks of deoxyribonucleic acids (DNA). The linear biopolymers are formed by the linear connection of the (deoxy-) nucleotides via phosphodiester linkages. The (deoxy-) nucleotides differ from each other by the base part: adenine (A), guanine (G), cytidine (C) and thymine (T). In the case of RNA, thymine is replaced by uracil (U).

in $3000 \times 56\,250 = 1.7 \times 10^8$ antibody specificities, whereby this value in practice is far surpassed by additional variation options not described here. This is an impressive demonstration of the fact that the combination of different elements to give new, diverse ensembles represents an elegant possibility, whenever variety is required.

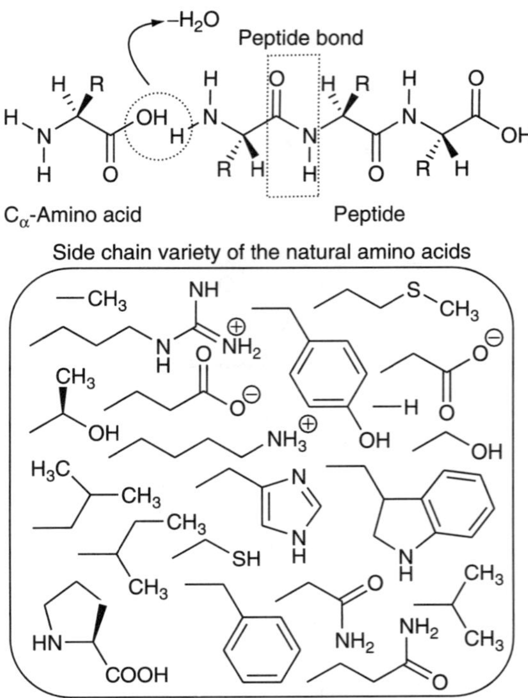

Figure 1.2 Amino acids are the building blocks of peptides and proteins. The linear biopolymers are formed by the polycondensation of the amino acids under formation of peptide or amide bonds. Twenty different amino acids exist in nature. This is sufficient to produce an inconceivable variety of peptides and proteins.

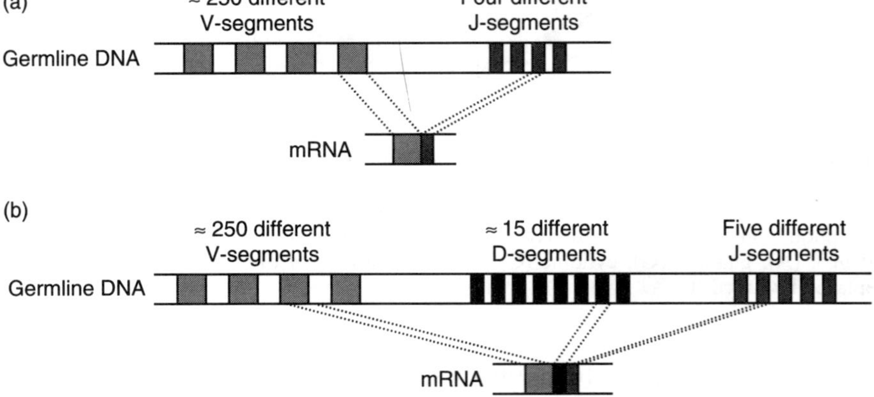

Figure 1.3 Antibody variety through combination of different gene segments. (a) The variable region of the light chain (κ-L-chain) is encoded by combination of V- and D-segments. (b) The variable region of the heavy chain (H-chain) is coded by combination of V- and J-segments as well as additionally by D-segments.

Sugar molecules are further simple building blocks of our bodies. However, in contrast to the above described building blocks, they have one additional feature: they can be connected to each other in different ways. For example, one finds glucose units in nature linked to each other by linear α-1,4- and β-1,4-glycosidic bonds; however, they are also often branched by α-1,6- and more rarely by α-1,2- or α-1,3-glycosidic bonds (Figure 1.4). Thus sugars not only enable the variation of the simple building blocks themselves, but also the variation of the type of connection.

One can easily imagine that there are nearly an infinite number of different structural elements that vary slightly and that can be used to generate polymers. In plastics research this has been happening for a long time, whereby the monomers are predominantly condensed to homopolymers such as polypropylene or polyamides (Figure 1.5).

In analogy to the biopolymers, from a combinatorial point of view, it was therefore natural to synthesize heteropolymers, for example oligo(*N*-substituted) glycines (so-called peptoids), oligocarbamates or oligoureas by linking different monomers (Figure 1.5).

However, the potential of combinatorics is not limited to polymers – on the contrary. Since every synthetically accessible molecule is built of different segments, it is obvious to vary the structural elements here as well, combine the residues and not only perform individual syntheses, but by using different building blocks, generate a whole set of molecules.

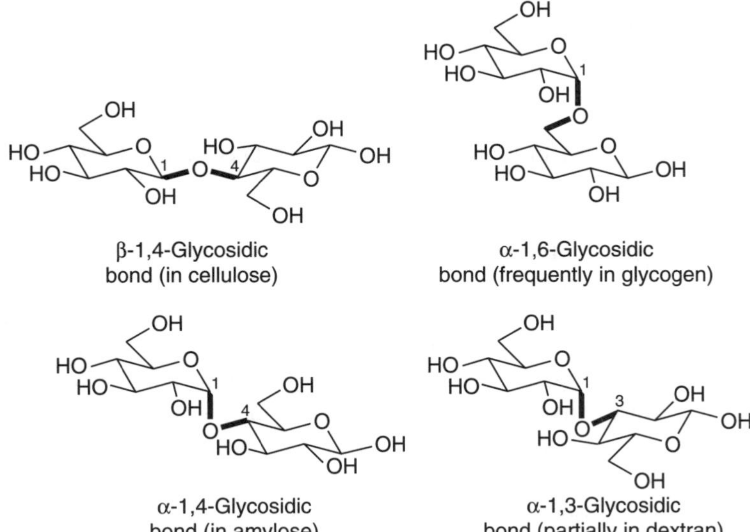

Figure 1.4 Different connection possibilities of sugar building blocks using the example of glucose.

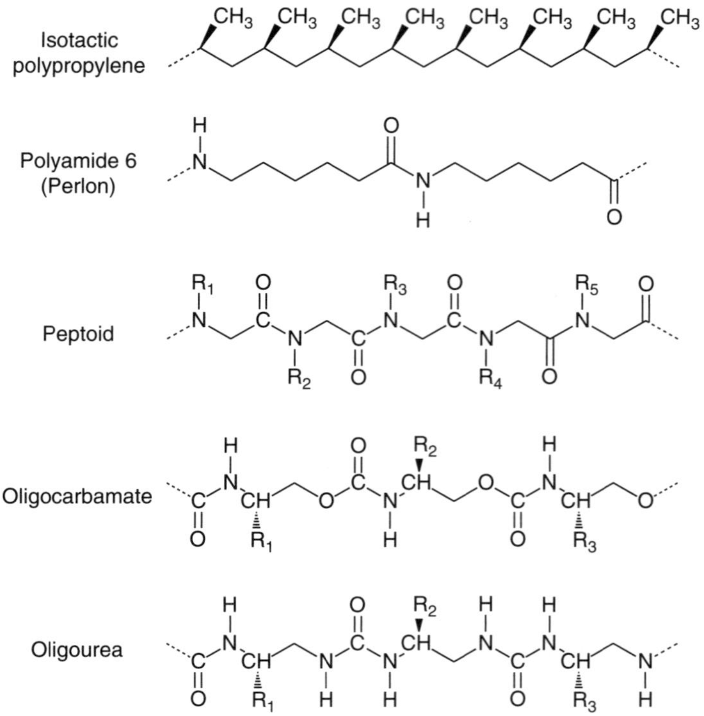

Figure 1.5 Examples of homopolymers (polymers made of identical building blocks), such as polypropylene and polyamide, as well as examples of combinatorial heteropolymers (polymers made of varied building blocks), such as peptoids, oligocarbamates and oligoureas.

1.3 THE SYNTHESIS OF MIXTURES AND THE MIXTURE OF SYNTHESES

Whenever we speak of combinatorial approaches, we mean the synthesis, examination and/or screening of hundreds to millions of more or less similar molecules. These are all manufactured in accordance with the same procedure, however, they differ by the arrangement of the building blocks or by the individual building blocks themselves. In the last few years, two principally different procedures have established themselves in this regard: parallel synthesis, which generates so-called arrays, and the synthesis of mixtures, also called pools or libraries (Figure 1.6).

Parallel syntheses usually run polymer bound. This makes possible the automation of the procedure with the help of robots, in turn making possible the synthesis of hundred to thousands of compounds. Parallel removal and addition of solvents with pipetting robots, parallel stirring, shaking, heating and

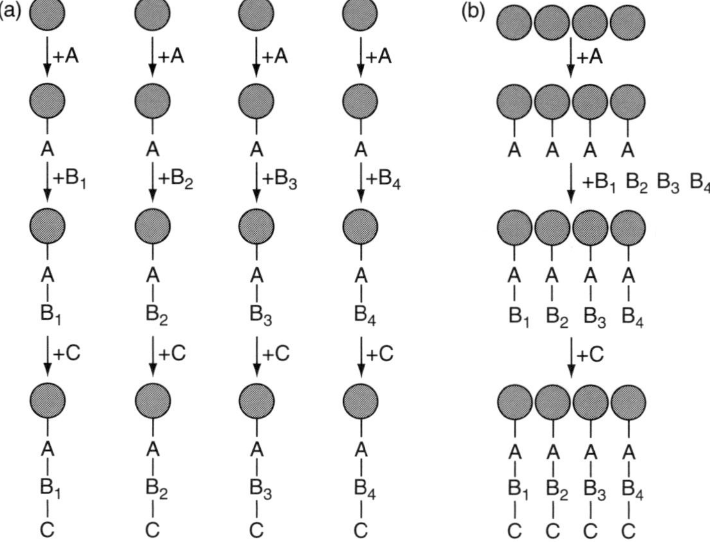

Figure 1.6 Comparison of the parallel synthesis, which generates so-called arrays (a), with the synthesis of mixtures, also called pools or libraries (b).

cooling is quite conceivable. On the other hand, parallel distilling, evaporation, crystallizing and sublimation for many slightly different compounds is difficult to realize and certainly cannot be generalized.

In contrast to parallel synthesis, no expenditures for apparatuses are necessary in the synthesis of mixtures. In a reaction which requires the connection of building blocks $A \rightarrow B \rightarrow C$, now not only is each building block added once, but rather, for example, in place of B, a mixture of four similar building blocks B_1, B_2, B_3, and B_4. After completion of the synthesis, we get a mixture of four products, namely AB_1C, AB_2C, AB_3C, and AB_4C, whereby B is described as the so-called 'mixed' position. If, in addition, we vary building block C through C_1, C_2, C_3, C_4, and C_5, the product mixture contains $4 \times 5 = 20$ different individual compounds. Generally the number of individual compounds of the mixture can be determined by multiplication of the building blocks used at each position, i.e. $A^x \times B^y \times C^z$. If the number of building blocks at each position is equal, for example, 20 amino acids, then the number of individual compounds in the mixture results from raising the building blocks *n* to the number of 'mixed' positions (Table 1.1). A mixture of tetrapeptides or longer peptides with four 'mixed' positions would, accordingly, contain $20^4 = 160\,000$ individual compounds. The 'mixed' positions are ordinarily described as X, whereby it is necessary to define out of which building blocks *n* each 'mixed' position X is composed. For example, the pentapeptide library YGXXL for optimization of encephalins is made up of 192 = 381 individual peptides. Two 'mixed' positions occur, and *n* encompasses in this case 19 natural

Table 1.1 Variation possibilities for polymers with x variable chain links, i.e. 'mixed' positions.

	$n^a = 2$	3	4	5	10	20	...	n
$x = 2$	4	9	16	25	100	40	...	n^2
3	8	27	64	125	1000	8000	...	n^3
4	16	81	256	625	10000	16000	...	n^4
5	32	243	1024	3125	100000	320000	...	n^5
6	64	729	4096	15625	10^6	6.4^7	...	n^6
7	128	2187	16384	78125	10^7	1.28^9	...	n^7
8	256	6561	65536	390625	10^8	2.56^{10}	...	n^8
⋮	⋮	⋮	⋮	⋮	⋮	⋮	⋮	⋮
x	2^x	3^x	4^x	5^x	10^x	20^x	...	n^x

[a] n = number of building blocks used per chain link.

amino acids, since cysteine was omitted due to its dimerization tendency, or oxidation sensitivity, respectively.

1.4 FROM PEPTIDE TO NONPEPTIDE

The origin of all chemical-combinatorial synthesis methods lies in peptide chemistry (Figure 1.7). Since the introduction of solid phase synthesis by Merrifield in 1963, peptides have been synthesized on solid supports, that is, the *C*-terminus of the peptide is anchored during the synthesis to a polymer, for example to modified polystyrene. All further building blocks are successively attached in the desired sequence. At the end the product is cleaved from the solid support, so that the peptide can be analyzed and purified. To be able to completely exploit the advantages of this very simple principle, a special chemistry was developed over the last 30 years of synthesis experience: suitable protecting groups, anchor groups to the polymer, activation techniques, solvents – everything possible was optimized and adapted for the special purpose of synthesis on the polymeric support. Even handling of the polymer itself requires a certain experience, which is not required with reactions in solution. For example, the polymeric support must be swollen, the reactions are diffusion controlled, and parameters such as pore size, uniformity, anchor selection and polymer quality influence the reactions. For these reasons, it is understandable that this knowledge and experience served as a foundation even in the widespread applications of combinatorial methods in the field of classic organic chemistry and still plays an important role. For example, the polymeric supports, protecting groups and anchors common in peptide synthesis are still most frequently employed.

The first mixture syntheses were carried out on polymeric supports with mixtures of peptides, the first large libraries were peptide libraries and after the automated, multiple peptide synthesis, one of the first array-synthesis was

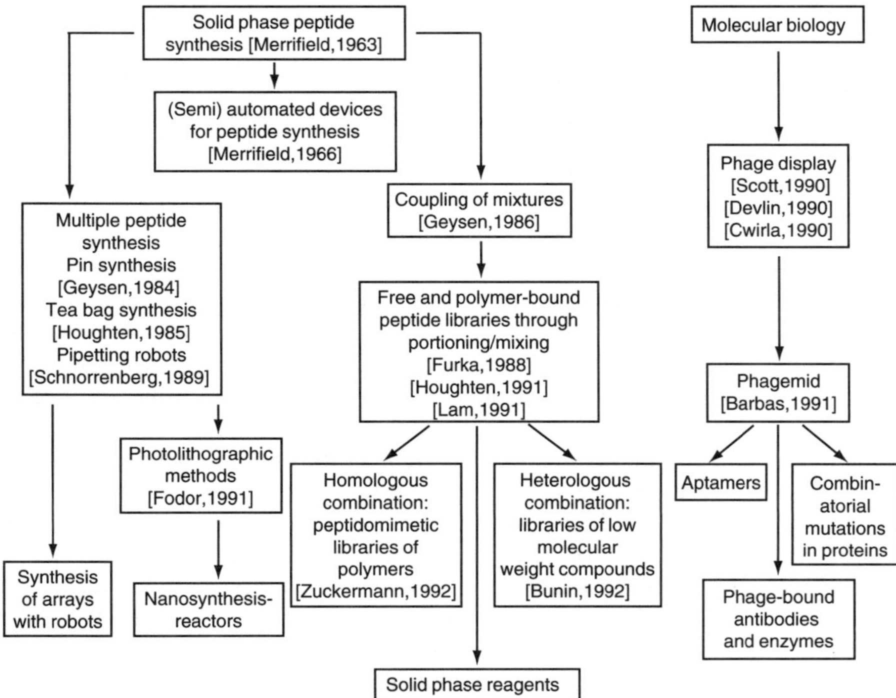

Figure 1.7 Historical overview of the developmental history of combinatorial methods in chemistry and biology.

the synthesis of peptides on chips by means of a photolithographic process (see Figure 1.7).

However, peptides have their strength in basic research, and are therefore at best suitable as lead structures and are of limited use as active compounds. As endogenous compounds they are often degraded too quickly to be used directly as pharmaceuticals. After the discovery of the concept of combinatorics in the development of active compounds, various concepts were and are being experimented with on the way to nonpeptides. Peptide libraries were synthesized with nonproteinogenous amino acids, D-amino acids were used or the peptides were modified (acylated, alkylated) after the synthesis. *N*-Substituted glycines, urea derivatives and many other structural elements were used for the synthesis of biopolymer mimetics in place of amino acids. Further, there were successful experiments in performing organic syntheses on polymeric supports and implementing combinatorial approaches based on these syntheses. Moreover, the automation that had been optimized within the scope of multiple peptide syntheses was transferred to other organic synthesis methods and attempted to automate not only the filtration and resuspension, but also the stirring, extracting and decanting.

1.5 SYNTHESIS IN SOLUTION OR ON THE SOLID SUPPORT

Development of the first synthesis automats was only possible through the step of the classic synthesis in solution to the solid phase. The polymer allows the standardization of the synthesis conditions. While it is true that individual syntheses may require different conditions, through the use of high excesses that are only possible through the working on polymers (otherwise the product would be very difficult to purify), frequently the reaction can be pushed in the desired direction. For many classic organic syntheses, however, up to now synthesis on the polymer was not possible or had not been developed: One requires a suitable anchor which then, and only then, releases the product at the end of the synthesis through suitable conditions. While the synthesis of a peptide, DNA segment or polycarbamate always requires the same set of reaction conditions the synthesis of a more complex molecule can require a multitude of reaction conditions, such as high temperatures or an inert atmosphere. In addition, not every molecule has a suitable functional group that is present at the beginning of the synthesis and remains unmodified to the end, since carboxylic acids and their amides were predominantly anchored on the polymer.

Thus, organic chemistry has developed in two directions with regard to combinatorial synthesis possibilities over the past few years. In one, the classic synthesis routes were varied for adaptation of the synthesis to the solid phase, which among others, resulted in development of new anchors for attachment of further functional groups such as alcohols, amines and aldehydes. In the other, classic synthesis routes were retained, whereby one only attempted to replace or circumvent processes that are difficult to automate: extraction instead of distillation, use of solvent mixtures to obtain as little precipitation as possible, or with opposite requirements, intentional precipitation. Naturally, corresponding automated synthesis lines require extremely complex robot systems, to be able to perform the manifold processes correctly.

The polymer-bound reagents and scavengers represent a third route. In this approach, the product is no longer bound to the polymer, but instead, the reagents are immobilized or byproducts get immobilized. The resulting carboxylic acid can be removed from the reaction solution by an amine resin, the resulting amine can be removed by an active ester resin. Couplings in solution are performed with the help of polymer-bound carbodiimide, racemization-free syntheses are performed with the help of polymer-bound 1-hydroxybenzotriazole (HOBt). The advantage is obvious: excess of reagents and scavenged byproducts are filtered off, the product remains in solution.

1.6 AUTOMATION: FROM MILLIGRAM TO MICROGRAM

While the first synthesis automats supplied gram amounts, in the past 10 years the trend has gone towards smaller and smaller compound amounts. How is this possible? And why is this of interest? Answering the second question is easy: If a compound is not active, which applies as a rule for at least 999 of 1000 intentionally manufactured compounds, it is not economically practical to synthesize much more of it than is required for the testing. The compound will definitely be kept for future applications, but for this purpose usually relatively small amounts suffice. On the other hand, if a compound is very active, relatively small amounts likewise suffice for this finding. In such cases, very large amounts are then required for extensive experiments hence the compound has to be resynthesized anyway. The reduction in compound and thus naturally also the reduction in cost are made possible by more effective test methods, with which the first question can also be answered. Instead of animal experiments, first studies are performed *in vitro* on cells, cell parts (for example, cell membranes), isolated enzymes or bacteria. While mg amounts were necessary per animal, a few μg are sufficient for the tests in the test tube. Effective distribution and robot-supported high-throughput screening allow the examination of the smallest compound amounts in shorter and shorter periods of time. However, in the future development will also continue in this area: Precise pipetting of nanoliter amounts will be routine and microtesting systems will be developed. Syntheses are to run on chips, capillaries could serve as reaction vessels. However, it is all too obvious that for this to occur, biologists, chemists, engineers, computer specialists and material scientists still have a number of open questions to resolve!

FURTHER READING

Abelson, J. N. (ed.) (1996). Combinatorial chemistry. *Methods in Enzymology*, vol. 267, Academic Press, San Diego.

Bunin, B. A. (1998). *The Combinatorial Index*, Academic Press, San Diego.

Chapman, K. T. Joyce, G. F. and Still, W. C. (eds) (1997). Current opinion in chemical biology. *Combinatorial Chemistry*, vol. 1, Current Biology Ltd., London.

Dörwald, F. Z. (2000). *Organic Synthesis on Solid Phase: Supports, Linkers, Reactions*, Wiley-VCH, Weinheim.

Famulok, M., Winnacker, E. L. and Wong, C. H. (eds) (1999). Combinatorial chemistry in biology. *Current Topics in Microbiology and Immunology*, vol. 243, Springer, Berlin.

Gordon, E. M. and Kerwin, J. F. J. (eds) (1998). *Combinatorial Chemistry and Molecular Diversity in Drug Discovery*, John Wiley & Sons, New York.

Jung, G. (ed.) (1996). *Combinatorial peptide and Nonpeptide Libraries – a Handbook*, Wiley-VCH, Weinheim.

Jung, G. (ed.) (1999). *Combinatorial Chemistry: Synthesis, Analysis, Screening*, Wiley – VCH, Weinheim.

Jung, G. and Beck-Sickinger, A. G. (1992). Multiple peptide-synthesis methods and their applications. *Angew. Chem Int. Ed. Engl.* **31**, 367–383.

Miertus, S. and Fassina, G. (eds) (1999). *Combinatorial Chemistry and Technology: Principles, Methods and Applications*, Marcel Dekker, New York.

Seneci, P. (2000). *Solid Phase Synthesis and Combinatorial Technologies*, John Wiley & Sons, New York.

Terrett, N. K. (1998). *Combinatorial Chemistry*, Oxford University Press, New York.
Wilson, S. R. and Czarnik, W. (eds) (1997). *Combinatorial Chemistry; Synthesis and Application*, John Wiley & Sons, New York.

2 Peptide Libraries – How it all Began . . .

2.1 SOLID PHASE PEPTIDE SYNTHESIS

In 1963 Merrifield introduced the principle of the synthesis of peptides on a solid matrix [Merrifield, 1963], through which peptide chemistry, up to this point rather inefficient, was revolutionized. The purification problems which inevitably resulted during the synthesis of peptides, which – if at all possible – could only be solved with great expenditure of time and technology, were elegantly solved in one fell swoop. The great, general significance of this completely new approach was reflected in the awarding of the Nobel Prize for Chemistry to Merrifield in 1984 [Merrifield, 1985]. The principle of solid phase synthesis also forms the foundation for synthesis automation and, closely related to this, for the later triumphant advance of combinatorial chemistry, for which reason we shall go into more detail in the following.

2.1.1 SOLID SUPPORTS

In general, polymers in the form of small granules (also called 'beads') with a diameter of 10–750 μm serve as solid supports (also called 'resin'). As one of their most important properties, the polymers must have good swelling characteristics, since the free diffusion of the reagents and solvent molecules in the interior of the granules must be guaranteed and the free unfolding of the polymer chains is the basic requirement for the smooth progression of reactions without steric hindrance. In addition, the polymers should also be as chemically inert as possible under the given reaction conditions and should have a certain mechanical stability. Further, it must be possible to covalently link the peptide to be synthesized with the polymer through a so-called linker or anchor (see Chapter 2, Section 1.2). In the normal case this linkage should be reversible. The number of anchors per resin bead (in this context one speaks of the 'loading') must be selected in such a way that intermolecular interactions between the growing peptide chains are avoided as far as possible to prevent incomplete coupling reactions. In general loadings of 0.2–1.2 mmol (anchors) per gram (resin) are common, whereby this value depends both on the resin type as well as on the size of the molecule to be synthesized. Table 2.1 lists some examples of the most common solid supports.

Table 2.1 Overview of the most common solid supports used in solid phase organic synthesis (also see Winter [1996]).

Polymer type	Description
Polystyrene resins	Polystyrene cross-linked with divinylbenzene (usually 1%)
TentaGel resins	Copolymer based on polystyrene, with grafted polyethyleneglycol (PEG)
PEGA resins	Bis-2-acrylamidoprop-1-yl polyethyleneglycol-acrylamido-prop-1-yl[2-aminoprop-1-yl] polyethyleneglycol dimethyl acrylamide copolymer
PolyHIPE resins	Copolymer based on polystyrene with grafted polydimethylacrylamide; HIPE = high internal phase emulsion
Pepsyn K resins	Polydimethylacrylamide, which is polymerized in an inorganic kieselguhr matrix
CP materials	Controlled pore glass (CPG) and controlled pore ceramics (CPC) are inorganic materials with defined pore size

Polystyrene Resins

Cross-linked polystyrene, which was already used by Merrifield, is still one of the most common solid supports today. The physicochemical properties of this gel-like material depend heavily on the degree of cross-linking of the polystyrene. The addition of 1% divinylbenzene (DVB) as a cross-linker has proven to be an optimum compromise between good swelling property (low cross-linking) and mechanical stability (high cross-linking). The resulting resin has good swelling properties in many different solvents, especially in the solvents mainly used for solid phase peptide synthesis, *N,N*-dimethylformamide (DMF) and dichloromethane (DCM) (see Table 2.2). Moreover, it is stable enough to

Table 2.2 Swelling volume in ml/g of polystyrene, which was cross-linked with 1% divinylbenzene (dry volume = 1.6 ml/g) and TentaGel resin (dry volume = 1.7 ml/g) following Rapp [1996].

Solvent	Polystyrene resin	TentaGel resin
Water	—	4.25
Methanol	1.6	4.25
Ethanol	1.68	2.1
Dichloromethane	8.3	5.1
Toluene	8.5	5.3
Dimethylformamide	5.6	5.4
Acetonitrile	3.2	5.1
Tetrahydrofuran	8.8	5.8
Dioxane	7.8	6.2
Diethyl ether	4.1	1.9

guarantee trouble-free handling and good filterability under usual peptide synthesis conditions. However, the hydrophobicity of the polymer is problematic. On the one hand, this prevents the use of polar solvents such as water, methanol or ethanol (see Table 2.2). On the other hand, in some cases it can result in an increase in the folding tendency of the hydrophilic peptide chains and in an induction of intermolecular peptide–peptide interactions because the peptides are striving to saturate their hydrogen bond donors and acceptors. As a result, in these cases one must reckon with insufficient coupling yields. If polystyrene resins are to be used for other solid phase syntheses, one should also keep in mind that the thermal stability of this polymer type is restricted (maximum temperature of approximately 125 °C).

Polyethylene Glycol-Containing Polymers

As an alternative to the hydrophobic polystyrene resins with their disadvantages, over the past few years hydrophilic copolymers of polystyrene and polyethylene glycol (PEG) were developed, which have gained great significance in peptide syntheses and especially in solid phase organic syntheses. The first and most important representative is the so-called TentaGel resin (Rapp Polymers, Tübingen, Germany) [Bayer and Rapp, 1986]. PEG chains are grafted to a matrix of polystyrene through ether bonds. The functional groups at the free termini of the PEG chains are used for the attachment of the compounds to be synthesized (see Figure 2.1). In general, PEG with a molecular weight of a few kDa is used for this purpose and a weight ratio of polystyrene to PEG of approximately 1–3 is strived for. Through this design principle the positive properties of both polymer types were able to be united: the stability of

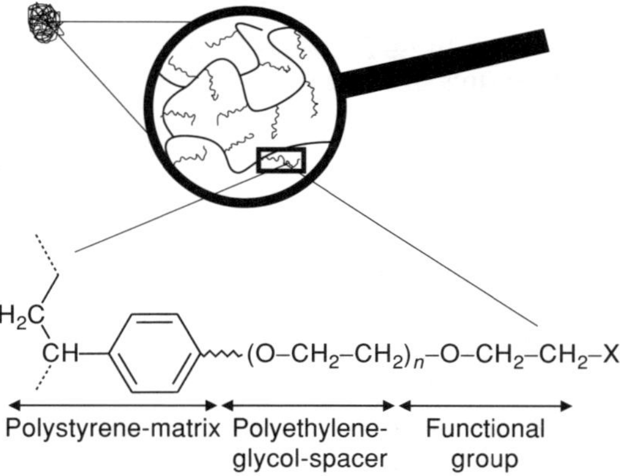

Figure 2.1 Structure of polystyrene–polyethylene glycol (PS–PEG) copolymers. Cross-linked PS serves as a base material, to which PEG tentacles were grafted. The synthesis takes place at the free end of each PEG unit (following Rapp [1996]).

the polystyrene and the extraordinary solubility and flexibility of the hydrophilic PEG [Rapp, 1996]. This even partially allows the testing of TentaGel-bound compounds in aqueous, that is biologically relevant assay systems (compare Table 2.2).

Meanwhile a number of similar resins have been developed, for example the so-called PEG–PS resin (PerSeptive, USA), in which the PEG molecules are linked with the polystyrene matrix through amide bonds. NovaGel resin (Novabiochem, Switzerland) is similar to TentaGel resin, but the anchors are not located at the end of the PEG chains, but rather directly at the polystyrene matrix. This design is based on the observation of a partial loss (so-called 'bleeding') of the grafted PEG chains, which in the case of TentaGel resin would lead to a loss of the synthesis products. However, through this reorientation of the anchors the microenvironment of the growing products becomes more hydrophobic.

As an alternative to PEG resins based on polystyrene, various approaches were undertaken to develop polymers directly on a PEG basis, since on the one hand, they improve the synthesis results for hydrophilic compounds, such as carbohydrates, and on the other, are even better suited for the direct testing of resin-bound compounds in aqueous environment. Since the PEG chains by themselves generally are somewhat too flexible, usually the copolymer approach was chosen: polyethylene glycol cross-linked oligostyrene (POEPS) resins consist of long PEG chains that are cross-linked with oligostyrene. In contrast to the polystyrene-based PEG resins the POEPS resins have a significantly greater PEG percentage with significantly greater pore size [Renil and Meldal, 1996]. Copolymers consisting of PEG and polyacrylamide (Polyethylene glycol dimethyl acrylamide (PEGA) resins) were likewise prepared through radical polymerization [Meldal, 1992]. Anionic or cationic polymerization with epichlorohydrin or oxetane yielded so-called POEPOP or SPOCC resins, which are able to withstand the most extreme synthesis and test conditions based on the inert character of the resulting polymers [Rademann *et al.*, 1999]. Figure 2.2 shows the chemical structures of some representatives of the PEG-based resins.

Polyacrylamide Polymers

In the 1970s and 1980s diverse resins were developed on a polyacrylamide basis [Atherton *et al.*, 1975; Arshady *et al.*, 1981], since these polymers have very good swelling properties in the polar, aprotic solvents commonly used for peptide synthesis (DMF, *N*-methylpyrrolidone (NMP)) and possess a certain similarity to the biopolymers to be synthesized. In addition, a series of copolymers were also introduced. Pepsyn K was among these, in which the polyacrylamide resin was polymerized to an inorganic kieselguhr matrix [Atherton *et al.*, 1981]. PolyHIPE (HIPE = high internal phase emulsion) is another representative of polyacrylamide copolymers and consists of a heavily branched

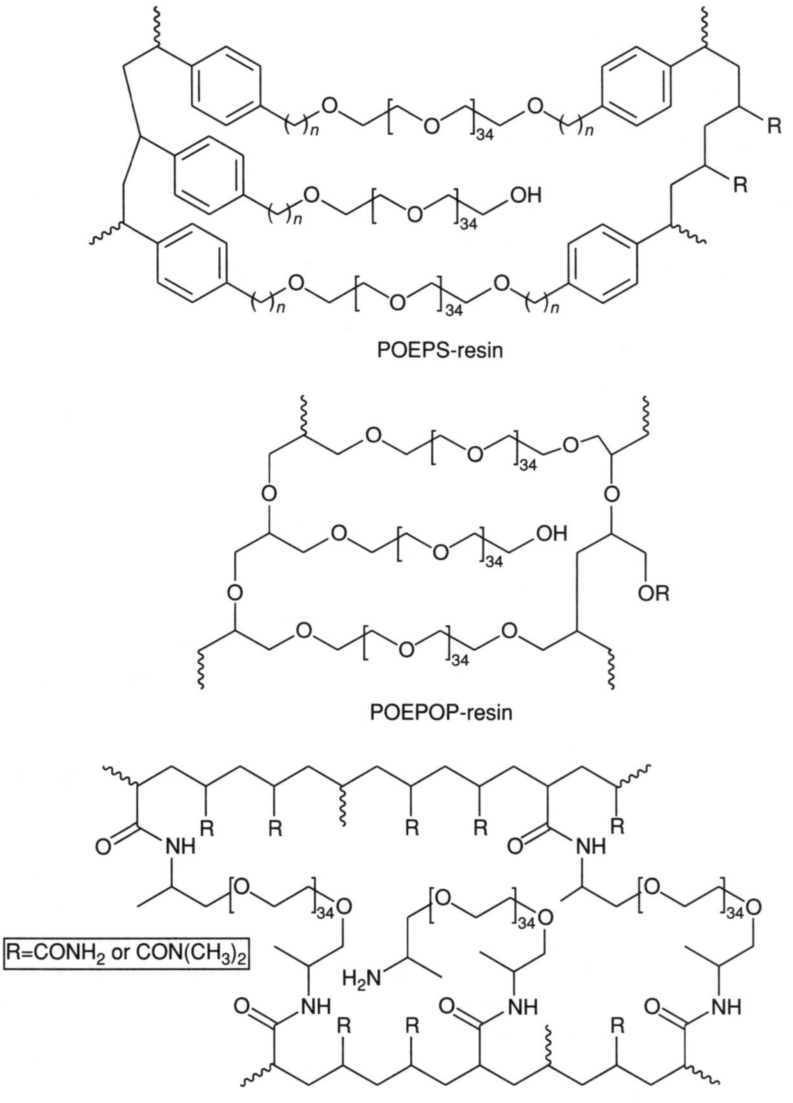

Figure 2.2 The chemical structure of PEG-based resins (following St Hilaire and Meldal [1999]).

macroporous polystyrene/divinylbenzene matrix, to which polydimethylacrylamide was grafted [Small and Sherrington, 1989]. By this means the physicochemical properties were able to be significantly improved in comparison to Pepsyn K, which is why the inhomogeneous inorganic matrix has practically disappeared from the market. The previous section has already covered PEGA resins as further representatives of the polyacrylamide copolymers.

Controlled Pore Glass or Ceramics (CPG, CPC)

If temperature stability or resistance to pressure are important, rigid inorganic materials such as glass or ceramics are alternatives. This is particularly the case with continuous flow syntheses. Through special manufacturing processes, in which pores with controllable sizes are introduced in the materials, the accessible surface for the reactants is enlarged, which allows loadings in the range of 0.2 mmol/g.

2.1.2 LINKERS/ANCHORS

For synthesis of the desired products, the solid supports must be equipped with appropriate groups, where the solid phase reaction can take place. These so-called linkers or anchors must be stable under the synthesis conditions, to prevent premature loss of the products. However, they must also possess a certain instability to allow the cleavage of the products from the solid support at the end of the synthesis. In addition, they must also guarantee the release of the products in their desired form, for example the cleavage of peptides as *C*-terminal amides or as free acids.

A classification of the linkers or anchors can be based on the cleavage conditions (cleavage with bases, acids, nucleophiles, light, enzymes, etc) or based on the resulting products (carboxylic acids, acid amides, esters, thioesters, alcohols, amines, etc). However, due to the diversity of the groups, an unambiguous assignment within the scope of both approaches is often difficult and confusing. In their reviews, though, several authors managed to present descriptive and comprehensive overviews of all the linkers/anchors [James, 1999; Warrass, 1999; Blackburn, 2000; Guillier *et al.*, 2000]. In the following we shall attempt to introduce the most important representatives, whereby the classification will be a synthesis of the product-oriented and the cleavage-oriented approach for reasons of clarity.

Carboxylic Acid Linkers

Merrifield, the inventor of solid phase peptide synthesis, functionalized his polystyrene-resin with chloromethyl groups to be able to anchor the *C*-termini of the growing peptide chains to the resin [Merrifield, 1963]. This so-called Merrifield resin was the standard support for the solid phase synthesis of peptide acids by the so-called *tert*-butyloxy carbonyl (Boc) strategy (see Section

2.1.4) and is still widely used for the synthesis of other carboxylic acid-containing compounds. Attachment of the carboxylic acids is generally achieved by heating the resin with the appropriate carboxylic acid cesium salt in DMF. Cleavage of the carboxylic acid is predominantly effected by treatment with hydrogen fluoride (HF) or trifluoromethyl sulfonic acid (TFMSA) (Figure 2.3). Today, one of the most commonly employed anchoring systems to peptide acids using the Boc strategy is the phenylacetamidomethyl (PAM) linker on polystyrene (see Section 2.1.4) depicted in Figure 2.3 which is also HF-labile but 100 times more stable during peptide synthesis than the Merrifield resin [Mitchell *et al.*, 1978].

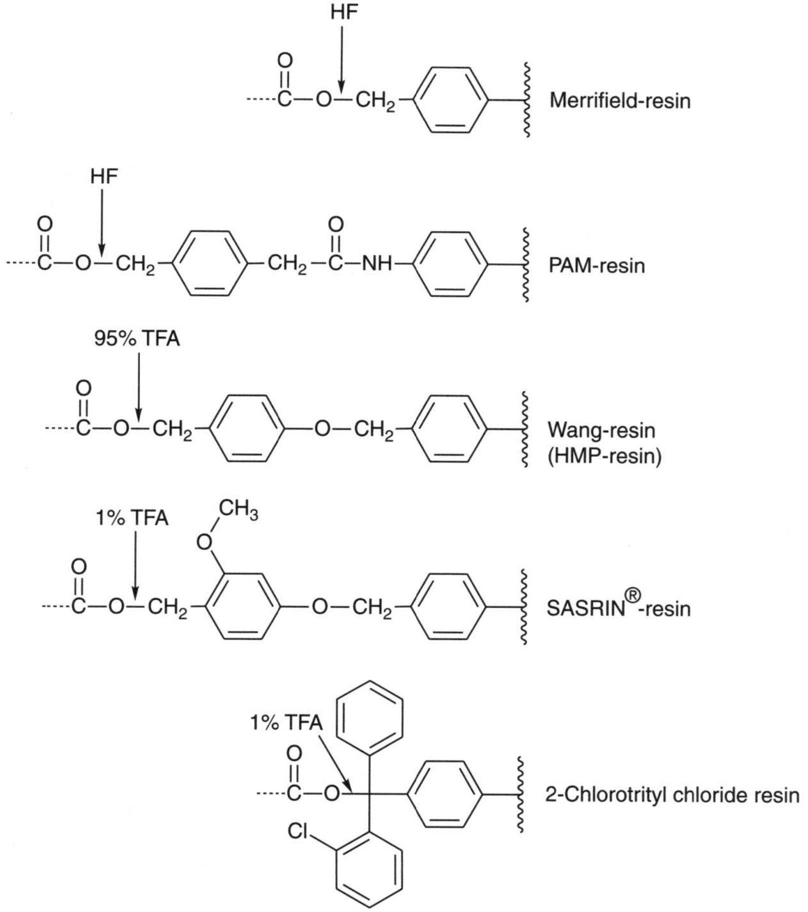

HF

$\cdots C(=O)-O-CH_2-C_6H_4-$ Merrifield-resin

HF

$\cdots C(=O)-O-CH_2-C_6H_4-CH_2-C(=O)-NH-C_6H_4-$ PAM-resin

95% TFA

$\cdots C(=O)-O-CH_2-C_6H_4-O-CH_2-C_6H_4-$ Wang-resin (HMP-resin)

1% TFA

$\cdots C(=O)-O-CH_2-C_6H_3(OCH_3)-O-CH_2-C_6H_4-$ SASRIN®-resin

1% TFA

$\cdots C(=O)-O-C(C_6H_5)(C_6H_4Cl)-C_6H_4-$ 2-Chlorotrityl chloride resin

Figure 2.3 Structures of the loaded anchors/resins that are typically used for the solid phase synthesis of peptide acids or other carboxylic acid-containing molecules. The cleavage conditions are indicated on top of the vertical arrows which mark the cleavage site.

Wang invented another type of linker based on 4-hydroxymethylphenol (HMP) [Wang, 1973]. The corresponding polystyrene-based HMP- or Wang-resin, respectively, became the standard support for the solid phase synthesis of peptide acids by the so-called 9-fluorenylmethoxycarbonyl (Fmoc) strategy (see Section 2.1.4), because the *para*-OCH_2-activated benzyl system allows the cleavage of the carboxylic acids with 95% trifluoroacetic acid (TFA), a much weaker acid than HF (Figure 2.3).

Adding a methoxy group to the Wang-linker leads to a super acid sensitive resin (acronym: SASRIN®) that releases the carboxylic acid even by treatment with 0.5–1% TFA (Figure 2.3) [Mergler *et al.*, 1988]. The acid lability is based on the electron-donating resonance effect of the substituent that stabilizes the cation which is formed during the acid catalysed cleavage. The resin is predominantly used for the preparation of partially protected peptide fragments by the Fmoc strategy (see Section 2.1.4).

The 2-chlorotrityl chloride linker is the most prominent representative of another group of acid-labile linkers, namely the trityl-based linkers [Barlos *et al.*, 1989a,b]. It can be cleaved by treatment with 0.5–1% TFA and can also be used for the preparation of partially protected peptide fragments (Figure 2.3). Its acid-lability is based on the remarkable stability of the bulky trityl cation, due to the delocalization of the positive charge within the aromatic system.

Carboxamide Linkers

The 4-methylbenzhydrylamine (MBHA) linker on polystyrene (Figure 2.4) is the standard support for the solid phase synthesis of peptide amides by the Boc strategy (see Section 2.1.4), because it releases the peptides as *C*-terminal carboxamides upon treatment of the resin with HF or TFMSA [Matsueda and Stewart, 1981].

The so-called Rink amide linker on polystyrene (Figure 2.4) became the most popular support for the solid phase synthesis of peptide amides by the Fmoc strategy (see Section 2.1.4) [Rink, 1987], but there exist some other supports with similar properties like the Knorr-resin [Bernatowicz *et al.*, 1989], and the peptide amide linker (PAL) resin [Albericio *et al.*, 1990]. They possess a greater acid-lability than the MBHA-linker which allows the cleavage of the carboxamides using 95% TFA. An even greater acid-lability is described for the Sieber resin that contains a xanthenyl-linker [Sieber, 1987]. In this case, the peptide amides can be released from the solid support using 1–5% TFA.

A more versatile approach is the conversion of a resin-bound carboxylic acid into a whole range of different carboxamides by nucleophilic attack in the final cleavage step. This can be achieved using the oxime linker which is also known as Kaiser's linker (Figure 2.4) [DeGrado and Kaiser, 1980]. Its aminolysis leads to the formation of primary carboxamides, but the synthesis of a variety of other products like hydrazides or esters is also feasible using the appropriate

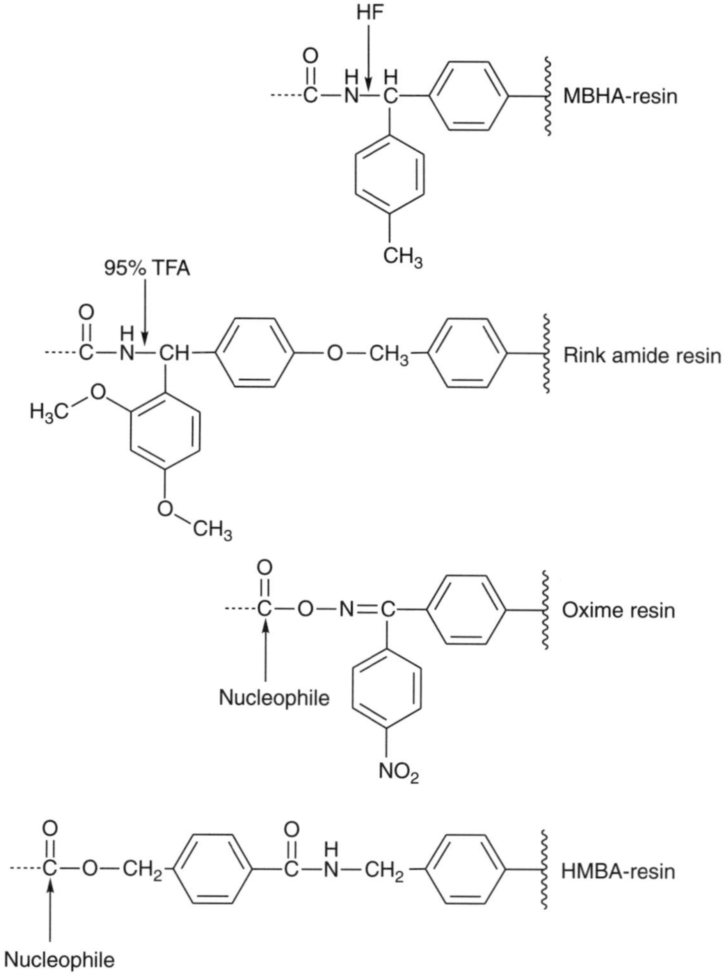

Figure 2.4 Structures of the loaded anchors/resins that are typically used for the solid phase synthesis of peptide amides or other carboxamide-containing molecules. The cleavage conditions are indicated on top of the vertical arrows which mark the cleavage site.

nucleophiles. However, the use of the oxime linker is limited to the synthesis of smaller peptides due to partial chain loss at each synthesis step. Another linker that allows the synthesis of a wide variety of *C*-terminal modified peptides including primary and secondary carboxamides, hydrazides, and esters by nucleophilic cleavage is Sheppard's 4-hydroxymethylbenzoic acid (HMBA) linker (Figure 2.4) [Sheppard and Williams, 1982].

A related, but more sophisticated approach is the synthesis of a peptide or organic molecule on a linker that is completely unreactive to the synthesis conditions, but can be activated by a subsequent chemical transformation to permit the cleavage of the compound by all kinds of nucleophiles. These linkers are called safety-catch linkers and are described in detail in a following section.

Amine Linkers

First attempts to link amines to the solid phase go back to the 1960s, when Letsinger *et al.* synthesized a small peptide starting at its *N*-terminus [Letsinger *et al.*, 1964]. Harsh cleavage conditions and the realization that peptide synthesis from the *C*- to the *N*-terminus is much more convenient stopped this approach. Within the past few years, the advent of combinatorial chemistry and solid phase organic synthesis led to the development of a wide variety of amine linkers and to procedures for the synthesis of amines using existing linkers. Carbamate linkers are the most popular class and can be cleaved with acids like TFA (Figure 2.5) [Marsh *et al.*, 1996]. The hydrazine-labile 4-acetyl-3,5-dioxo-1-methylcyclohexane carboxylic acid (ADCC) linker (Figure 2.5) is a good alternative, especially if the synthesis needs the use of strong acids, because the linker is stable to acidic conditions [Bannwarth *et al.*, 1996]. The well-known 2-chlorotrityl resin (see section on carboxylic acid linkers) was also used to attach primary amines and even secondary amines to the solid phase [Hoekstra *et al.*, 1997]. Tertiary amines can also be synthesized on the solid phase using the so-called REM resin (Figure 2.5), because the linker is

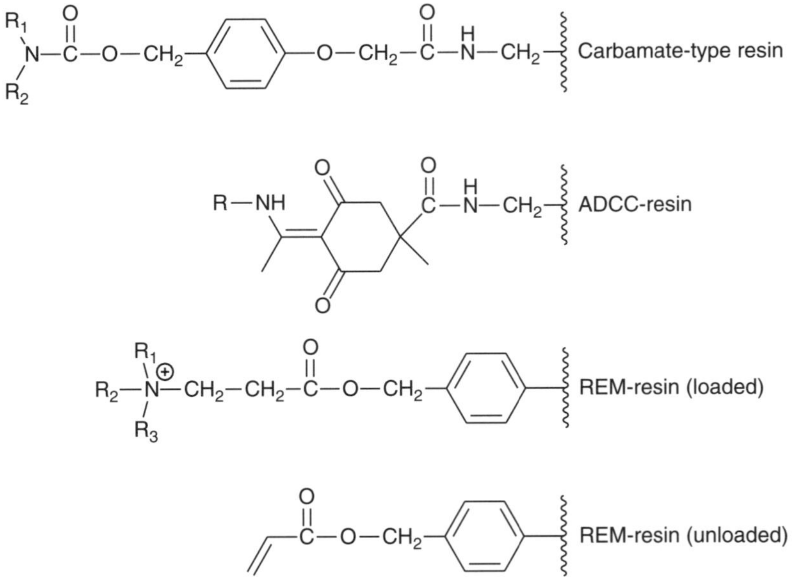

Carbamate-type resin

ADCC-resin

REM-resin (loaded)

REM-resin (unloaded)

Figure 2.5 Structures of the loaded (for REM-resin: and unloaded) anchors/resins that are typically used for the solid phase synthesis of amines.

regenerated at the end and a Michael addition is part of the synthesis [Morphy *et al.*, 1996].

Alcohol Linkers

Resins like Merrifield-, Wang-, and 2-chlorotrityl resin that are typically used for the synthesis of peptide acids can also be used for the synthesis of alcohols as shown by several groups [Kurth *et al.*, 1994; Wenschuh *et al.*, 1995; Alsina *et al.*, 1997]. A completely new approach for the attachment of hydroxyls to the resin is based on 3-hydroxymethyl-2,3-dihydro-4H-pyran, also known as tetrahydropyranyl (THP) linker (Figure 2.6) [Thompson and Ellman, 1994]. It can be used for all kinds of alcohols and is stable to strong bases, but can be cleaved by 95% TFA.

Traceless Linkers

Traceless linkers can be cleaved from the resin leaving no residual functionality. This enables the attachment of arenes and alkanes to a polymeric support and allows the combinatorial synthesis of new classes of low molecular weight organic compounds. Aryl silyl systems (Figure 2.7) were the first traceless linkers to be developed and are widely used for the synthesis of compounds containing aromatic hydrocarbons [Chenera *et al.*, 1995; Plunkett and Ellman, 1995]. The silicon–phenyl bond is sensitive to acids and other electrophiles and can undergo protodesilylation or halodesilylation reactions yielding unsubstituted phenyls or aryl halides, respectively. Sulfur- and selenium-based linkers are another important class of traceless linkers [Zaragoza, 2000]. For example, sulfide linkers can be used for aliphatic C–H-bond formation by oxidizing the sulfide to a sulfone and subsequent reductive cleavage with sodium amalgam (Figure 2.8) [Zhao *et al.*, 1997]. Additional information about traceless linkers can be gathered in an excellent review [Brase and Dahmen, 2000].

Safety-Catch Linkers

Safety-catch linkers are completely inert to the synthesis conditions, but have to be chemically transformed to allow the final liberation of the product from the

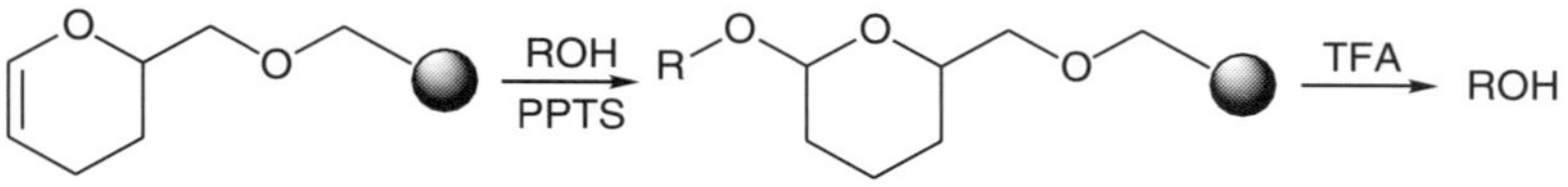

Figure 2.6 Loading and cleavage of the tetrahydropyranyl (THP) linker that is used for the solid phase synthesis of alcohols [Thompson and Ellman, 1994].

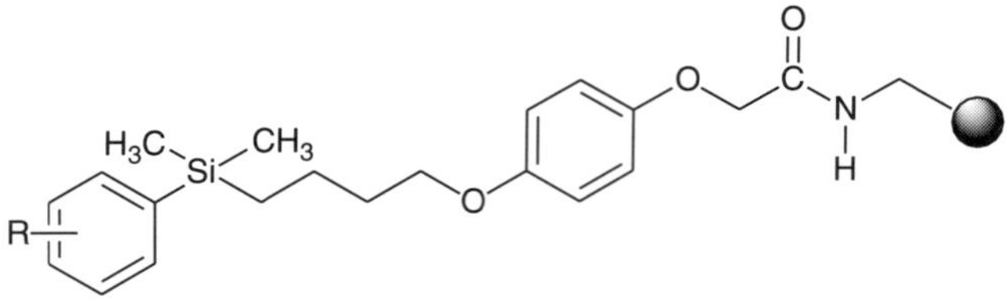

Figure 2.7 Traceless silyl linkers for the solid phase synthesis of aromatic compounds. Top: Ellman silyl linker [Plunkett and Ellman, 1995]; bottom: Veber silyl linker [Chenera *et al.*, 1995].

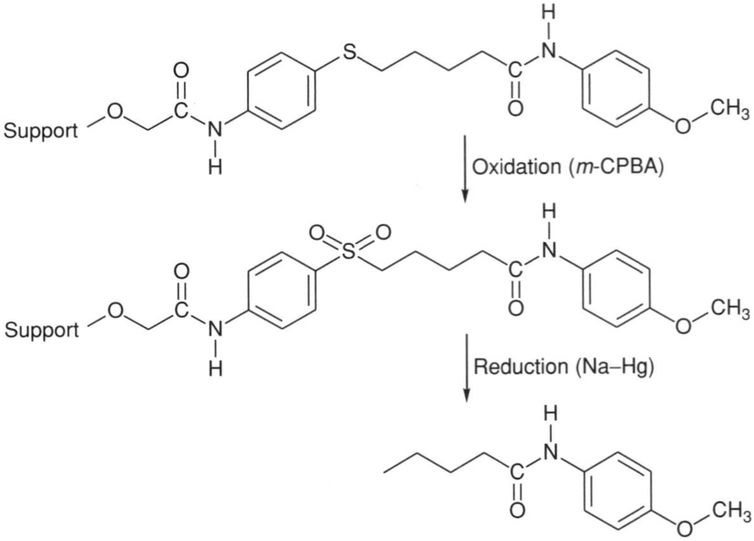

Figure 2.8 Traceless sulfide linker for aliphatic C–H-bond formation by an oxidation–reduction sequence [Zhao *et al.*, 1997].

solid phase. This is particularly useful for the synthesis of peptide *C*-terminal thioesters by Fmoc strategy. The thioesters are required for the so-called native chemical ligation of peptide or protein fragments. Thioesters are stable under acidic conditions and can be synthesized easily using Boc strategy, but are highly sensitive to nucleophilic bases which prevents their synthesis by standard Fmoc strategy (see Section 2.1.4). An acylsulfonamide safety-catch linker developed by Kenner *et al.* [Kenner *et al.*, 1971] and modified by Backes *et al.* [Backes *et al.*, 1996; Backes and Ellman, 1999] allows the synthesis of peptide thioesters by Fmoc strategy, because the thioester is formed after the synthesis of the peptide in the final cleavage step using a nucleophilic thiol (Figure 2.9) [Ingenito *et al.*, 1999; Shin *et al.*, 1999]. Since the final cleavage can be performed with a wide variety of nucleophiles, the linker is a versatile tool for the synthesis of carboxylic acids, carboxamides or hydrazides. The authors suggest the review by Patek and Lebl for detailed information about other safety-catch linkers and their use for the synthesis of nonpeptide compounds [Patek and Lebl, 1998].

Photocleavable Linkers

The liberation of compounds from the resin by light is a very elegant approach since the linkers are completely unreactive to common synthesis conditions and

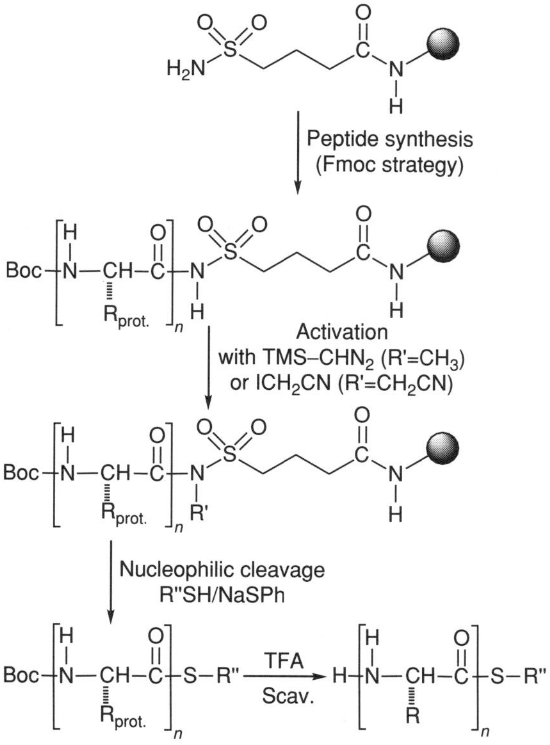

Figure 2.9 Use of an acylsulfonamide safety-catch linker for the synthesis of peptide *C*-terminal thioesters by Fmoc strategy according to Ingenito *et al.*, [1999] and Shin *et al.*, [1999].

the final cleavage just requires some scavengers due to its noninvasive nature. The most effective photolinkers were developed by Affymax and allow the release of carboxylic acids as well as carboxamides by ultraviolet light (UV) (Figure 2.10) [Holmes and Jones, 1995; Holmes, 1997].

Miscellaneous Linkers

Allyl-functionalized linkers like HYCRAM® (hydroxy-crotonyl-aminomethyl) [Kunz and Dombo, 1988] and HYCRON® (Figure 2.11) [Seitz and Kunz, 1997] offer a further strategy to release compounds (mostly carboxylic acids) from the solid phase: palladium catalysed cleavage of allyl esters.

Multifunctional resins like [acyl-4-(4-oxymethyl) phenylacetoxy-methyl]-3-nitrobenzoic acid (Pop) or acyl-2-[(oxymethyl)phenylacetoxyl]-propionyl (Pon) resins (Figure 2.12) possess multiple cleavage sites and allow the generation

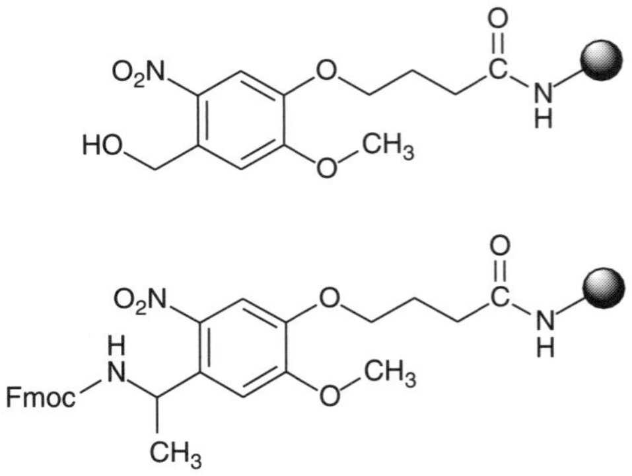

Figure 2.10 Photocleavable linkers for the solid phase synthesis of carboxylic acids (top) and carboxamides (bottom) according to Holmes and Jones [1995] and Holmes [1997].

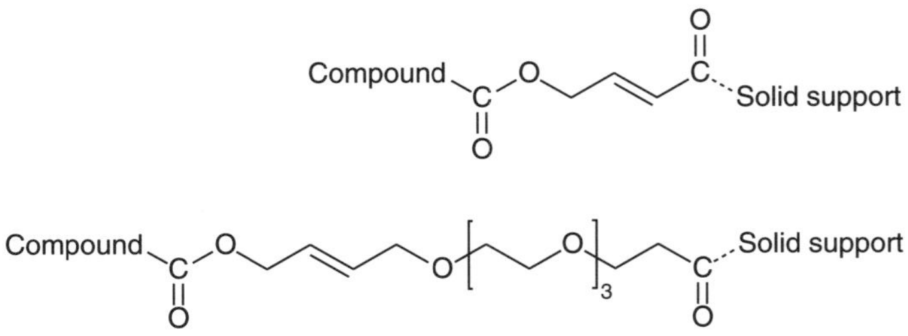

Figure 2.11 Allyl-functionalized anchors HYCRAM® (top) and HYCRON® (bottom), that can be cleaved by palladium-catalysis to yield carboxylic acids [Kunz and Dombo, 1988; Seitz and Kunz, 1997].

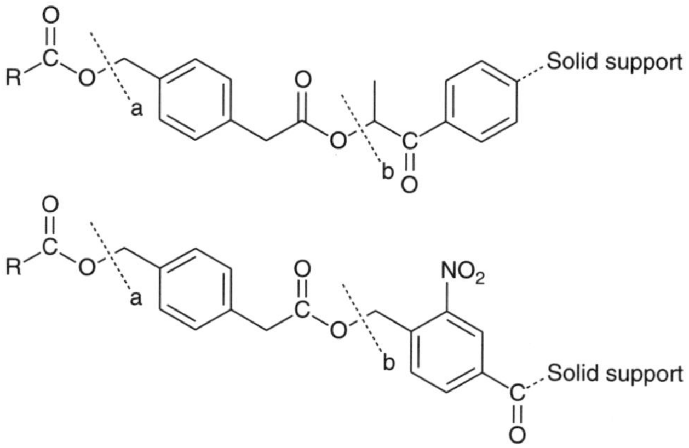

Figure 2.12 Multifunctional anchors Pop (top) and Pon (bottom), that have two different cleavage sites a (acid or hydrogenolysis) and b (photolysis) [Tam *et al.*, 1979, 1980].

of various end-groups by the use of different cleavage strategies [Tam *et al.*, 1979, 1980].

There are a few other examples for linkers that can be cleaved by enzymes – reduction, oxidation, and hydrogenolysis, respectively [Warrass, 1999]. They proved to be useful for certain applications, but lack general relevance.

2.1.3 SYNTHESIS PRINCIPLE

The synthesis of a peptide at solid supports is based on a relatively simple principle and normally occurs from the *C*-terminus to the *N*-terminus. First, in a preceding step the carboxyl group of the *C*-terminal amino acid of the desired peptide has to be attached to the resin through an anchor (see Section 2.1.2). To prevent side reactions, both the α-amino group as well as any functional group of the side chain of the amino acids has to be reversibly protected. To extend the peptide chain the cycle described in Figure 2.13 has to be run once for each amino acid and repeated until the peptide is completely built up. Finally the peptide is liberated from the support by cleavage of the anchor and the side chain protecting groups (PGs).

For a successful solid phase peptide synthesis the coupling reactions and the deprotections have to succeed quantitatively and putative side reactions (for example due to unstable side chain protecting groups) have to be suppressed as much as possible. One reason for this is that neither a purification nor a characterization of the resin bound intermediates is possible with conventional analytical methods. Another reason is that the peptidic byproducts often have such a high chemical and physical similarity to the cleaved product that its purification and characterization also presents a challenge.

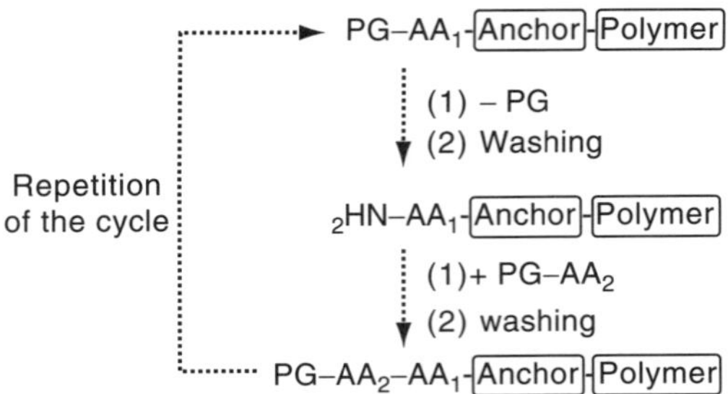

Figure 2.13 Schematic representation of solid phase peptide synthesis. The *C*-terminal amino acid (AA_1) is coupled to the polymer as an *N*-terminally protected derivative (PG–AA_1) via a cleavable anchor group. Next, the following cycle is repeated until the desired peptide is completely built up: cleavage of the *N*-terminal protecting group; separation of excess reagents and nonresin-bound byproducts, coupling of the next amino acid; again separation of excess reagents and nonresin-bound byproducts.

2.1.4 PROTECTING GROUP STRATEGIES

To guarantee an unequivocal progression of the coupling reactions, both the α-amino groups and the reactive side chains of the trifunctional amino acids have to be protected in reversible form. It has to be taken into account that the stabilities of side chain protecting groups and *N*-terminal protecting groups must be graduated in some manner, since the cleavage of the *N*-terminal protecting group, which occurs after each coupling step, may not influence the stability of the side chain protecting groups. For their part, the side chain protecting groups may not also be too stable, so that they can be cleaved at the end of the synthesis without damaging the actual peptide structure.

In practice, two protecting group strategies have proven successful and are used today in the overwhelming majority of solid phase peptide syntheses. We are referring to the so-called Boc/benzyl and the Fmoc/*t*Bu strategies.

The Boc/benzyl strategy (Figure 2.14) dates back to Merrifield and is based on the reversible blockage of the α-amino group of the amino acid used for chain elongation with the *tert*-butyloxycarbonyl (Boc) protecting group. This

Figure 2.14 Schematic representation of the Boc/benzyl strategy for solid phase peptide synthesis.

can be quantitatively removed with a 1:1 mixture of dichloromethane (DCM)/trifluoroacetic acid (TFA). The side chains of trifunctional amino acids on the other hand are protected mainly with protecting groups of the benzyl type, which are sufficiently stable under the conditions used for the removal of the Boc protecting group. The cleavage of these side chain protecting groups requires stronger acids like anhydrous hydrogen fluoride (HF) or trifluoromethanesulfonic acid (TFMSA). The substituted benzyl ester PAM or *N*-benzyl amide MBHA (see Section 2.1.2) have proven successful as anchors, since they have a comparable TFA stability and HF lability to the side chain protecting groups and the peptides can be released under the formation of native free acid or amide *C*-termini, respectively (Figures 2.3 and 2.4).

The Fmoc/*t*Bu method is a so-called 'orthogonal' protecting group strategy, because it is not based on a graded acid lability of the protecting groups as is the Boc/benzyl strategy, but rather on the base lability of the α-amino protecting group 9-fluorenylmethoxycarbonyl (Fmoc) and on the acid lability of the side chain protecting groups (mainly from the *tert*-butyl (*t*Bu) type) and the anchor (Figure 2.15). *p*-*O*-activated benzyl esters like Wang or amides like Rink amide are mainly used as anchors in the Fmoc/*t*Bu strategy (see Section 2.1.2, and Figures 2.3 and 2.4). The highly acid sensitive resins SASRIN® and 2-chlorotrityl allow the cleavage of fully protected peptides from the resin. This means that all side chain protecting groups remain on the peptide which is impossible within the scope of the normal Boc/benzyl strategy, due to the lacking orthogonality. This has made the Fmoc/*t*Bu strategy more and more attractive and important in the last few years, because in a subsequent step the fragments can be selectively condensed or modified in solution, through which it is possible, for example, to synthesize larger peptides or proteins chemically (so-called convergent strategies). Along with the more variable and more sparing cleavage conditions, the Fmoc/*t*Bu strategy also offers some advantages concerning technical aspects in comparison to the Boc/benzyl strategy: work with extremely dangerous gaseous hydrogen fluoride can be avoided and along with this it is unnecessary to buy the associated expensive Teflon equipment. Above all, this is of importance with regard to an automation of the solid phase synthesis.

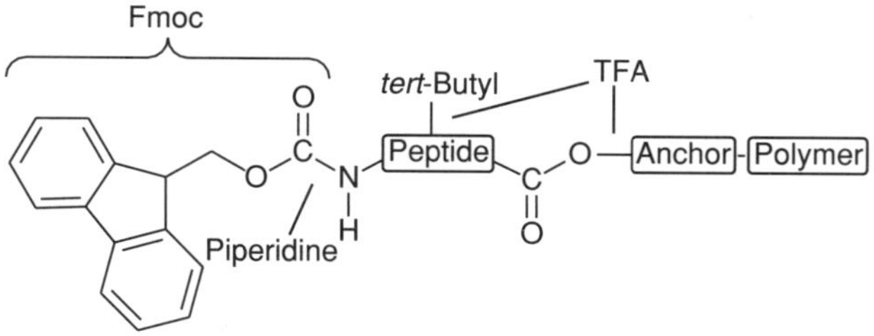

Figure 2.15 Schematic representation of the Fmoc/*t*Bu strategy for solid phase peptide synthesis.

It should also be pointed out that in addition to the anchor-variants mentioned in Section 2.2 of this chapter, a multitude of other protecting groups and anchors are available to the peptide chemist, which have orthogonal properties in the third dimension. In the case of the Fmoc/*t*Bu strategy, this means that they cannot be attacked by acids or bases and thus, for example, allow the selective side chain modification of trifunctional amino acids. In this context, the protecting groups of allyl type (Figure 2.16) offer interesting perspectives, since they can be cleaved under neutral conditions using a Pd catalyst. The protecting groups of 1-(4,4-dimethyl-2,6-dioxocyclohexylidene)ethyl (Dde) type also represent a good alternative. They can be selectively cleaved with 2% hydrazine in DCM, but are stable under acidic and partially under alkaline conditions (Figure 2.16) [Bycroft *et al.*, 1993]. Acetamidomethyl (Acm) protecting groups of cystein side chains can be cleaved selectively under oxidative conditions (for example with I_2) (Figure 2.16). Further, we should mention photolabile protecting groups (Figure 2.17), since they allow the selective deprotection of functional groups in an elegant manner without influencing other protecting groups (see section on photocleavable linkers in this chapter).

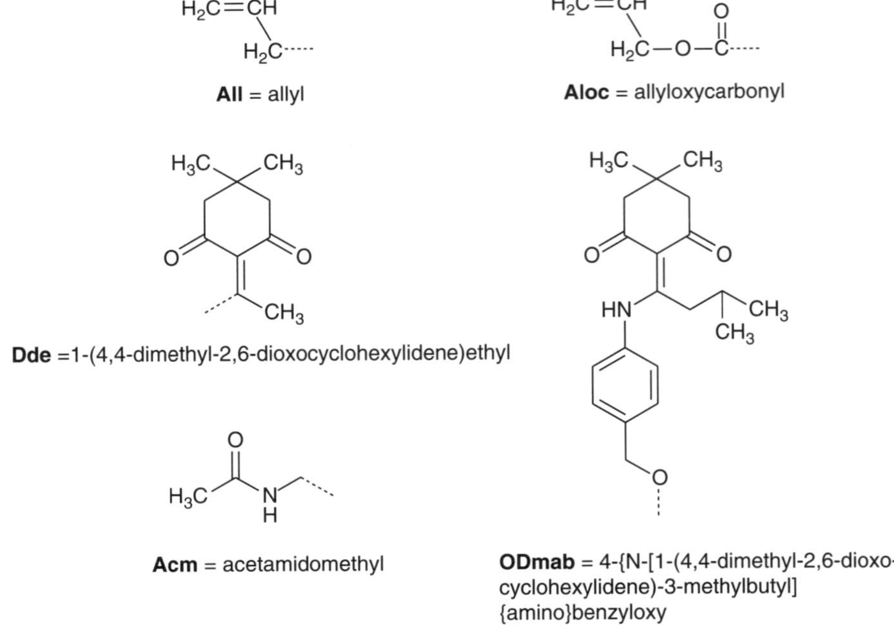

Figure 2.16 Examples of selectively removable side chain protecting groups for the Fmoc/*t*Bu strategy.

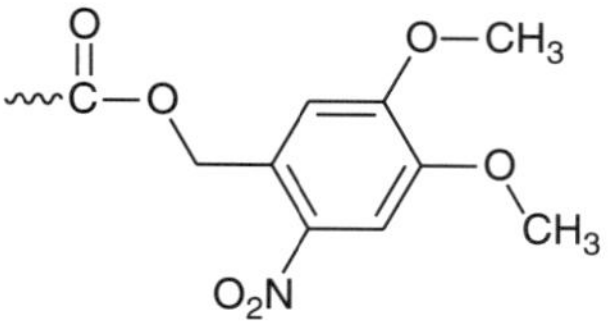

Figure 2.17 Photolabile amino protecting group 6-nitroveratryloxycarbonyl (Nvoc). Deprotection occurs by long-wave UV irradiation ($\lambda = 350$ nm).

2.1.5 ACTIVATION METHODS

The carboxyl group, which reacts sluggishly in and of itself, has to be converted into an activated derivative for an efficient reaction of the carboxyl group of the amino acid to be coupled with the α-amino group of the amino acid already bound to the resin. This takes place through the introduction of an electron-attracting group X that, on the one hand, increases the positive partial charge on the carboxyl C-atom and, on the other hand, acts as an excellent leaving group (Figure 2.18).

The various carbodiimide methods have special significance for solid phase peptide synthesis (independently of the protecting group strategy), particularly in combination with the use of additives such as 1-hydroxybenzotriazole (HOBt) or *N*-hydroxysuccinimide (NHS) (Figure 2.19).

Coupling reagents based on phosphonium salts also became very popular, but were replaced relatively quickly by the corresponding uronium salts due to the toxicity of the reaction products that arose (Figure 2.20). Further, acid halides created *in situ* are being increasingly used, because they can react with little steric hinderance. Furthermore, activated compounds of Fmoc-protected amino acids exist that are stable as crystalline solids. This has the advantage that these amino acid derivatives can be directly coupled without the addition of activating reagents and, in addition, are commercially available. The pentafluorophenyl esters and NHS esters depicted in Figure 2.18 belong to this group.

Using these optimized activation methods it has become possible in the meantime to realize virtually quantitative coupling yields with a relatively low excess of amino acid. Thus, efforts for purification and characterization of the products, as well as costs for the synthesis, could be effectively reduced.

2.2 MULTIPLE SOLID PHASE PEPTIDE SYNTHESIS (MPS)

In order to satisfy the rapidly increasing demand for synthetic peptides of various lengths and amino acid compositions, several groups, independently of one another, developed different concepts for the parallel – and thus

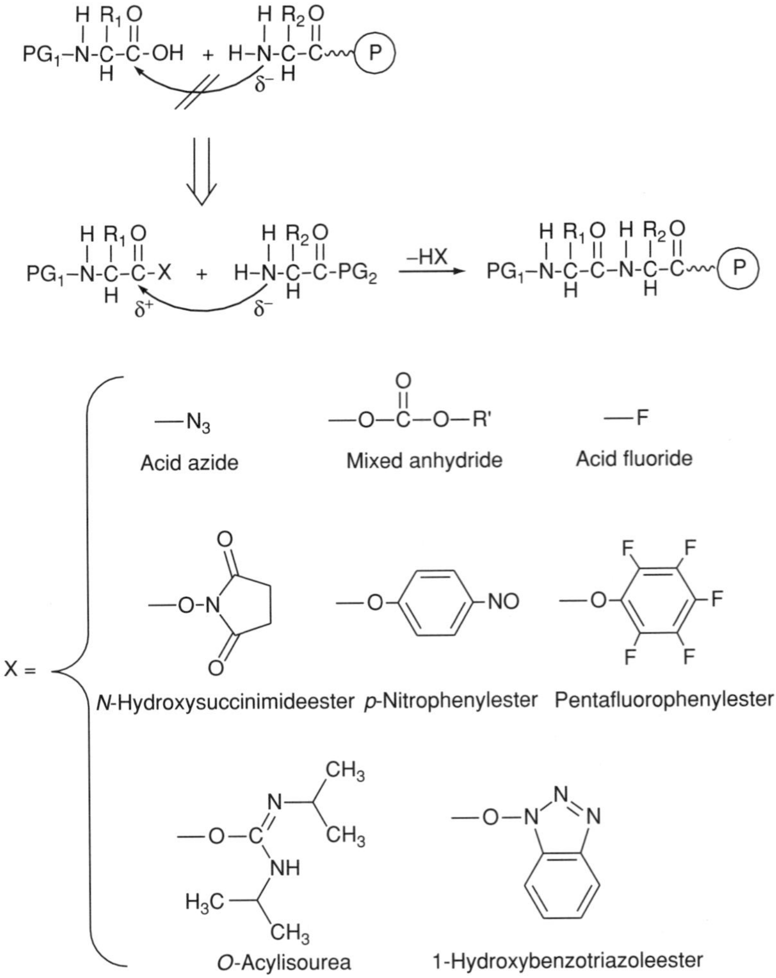

Figure 2.18 Activation of the carboxyl group to form a peptide bond. The carboxyl group has to be converted to a more reactive derivative in order to allow the nucleophilic attack of the amino group on the carboxy-*C* atom. This takes place through the introduction of an activating group X that has an electron-attracting effect and represents a good leaving group.

simultaneous – synthesis of peptides and therein started the age of combinatorial chemistry (Table 2.3). These so-called multiple peptide syntheses (MPS), which are also called simultaneous multiple peptide syntheses (SMPS), are all based on the solid phase peptide synthesis strategies mentioned above. Therefore, the same activation methods, protecting group strategies and polymeric supports are used. The various approaches only distinguish themselves from one another in their technical execution and, associated with that, in the number of possible peptides and the amounts of products obtained.

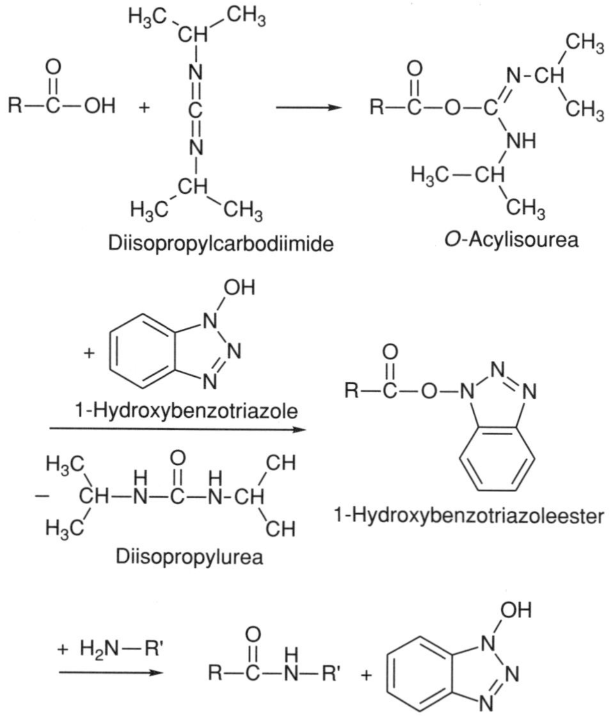

Figure 2.19 Mechanism of activation via the 1,3-diisopropylcarbodiimide (DIC)/1-hydroxybenzotriazole (HOBt) method. The *O*-acylisourea is formed first, but tends to rearrange and racemize. HOBt intercepts this intermediate which leads to the formation of an active ester, that then represents the actual reactive species (following Altmann and Mutter [1993]).

Table 2.3 Overview of basic methods of multiple solid phase peptide synthesis.

Author (year)	Polymer	Approach	Strategy	Reference
Geysen (1984)	Functionalized PE pins	Batch	Boc	[Geysen *et al.*, 1984]
Houghten (1985)	Polystyrene–divinylbenzene in tea bags	Batch	Boc	[Houghten, 1985]
Frank (1988)	Cellulose discs	Continuous flow	Fmoc	[Frank and Döring, 1988]
Krchňák (1989)	Polystyrene–divinylbenzene	Continuous flow	Boc	[Krchnák *et al.*, 1989]
Berg (1989)	Polystyrene–polyethylene films	Batch	Boc	[Berg *et al.*, 1989]
Eichler (1989)	Cellulose paper	Batch	Fmoc	[Eichler *et al.*, 1989]
Lebl (1989)	Cotton	Batch	Fmoc	[Lebl and Eichler, 1989]

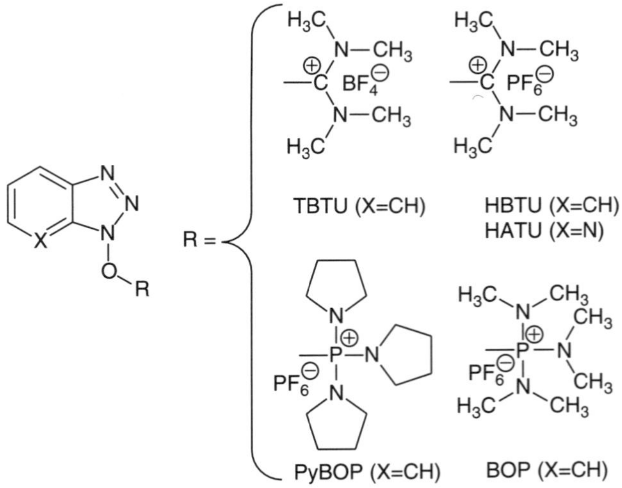

Figure 2.20 Recently developed coupling reagents based on phosphonium and uronium salts. The X-ray structure analysis of HATU and HBTU showed that these two reagents do not have the uronium salt structures proposed in the literature for over two decades but rather are guanidinium *N*-oxides.

2.2.1 PEPTIDE SYNTHESIS ON PINS

In 1984, Geysen *et al.* were the first to publish a paper on the parallel synthesis of peptides on pins and can therefore be viewed as the actual founders of combinatorial chemistry [Geysen *et al.*, 1984].

As schematically presented in Figure 2.21, amino-functionalized polyethylene pins (diameter 4 mm, length 40 mm) are used as carriers. Ninety-six of these pins are fixed on a block in eight rows of 12 pins each, so that they fit exactly into the wells of microtiter plates, as they are also used for enzyme-linked immunosorbent assays (ELISAs). The coupling of the amino acids takes place on these plates; the amounts and the distribution of the Fmoc amino acids to be coupled can be calculated by a computer program.

The polyethylene pins can neither swell nor shrink. Thus, it is difficult to remove adsorbed molecules. Intensive washing after every reaction step is therefore of crucial importance. This – and the deprotection steps – takes place in suitable tubs.

The specialty of this method lies in the fact that the peptides remain on the pins not just during, but also after the synthesis, and only the protecting groups are removed. An analysis, and thereby purity control of the synthesized peptides, is consequently impossible without further effort. Due to the low product amounts of maximally 300 nmol peptide/pin an attempt to analyze the peptides would have been extremely difficult anyway, taking into consideration the sensitivity of the analysis methods available at that time.

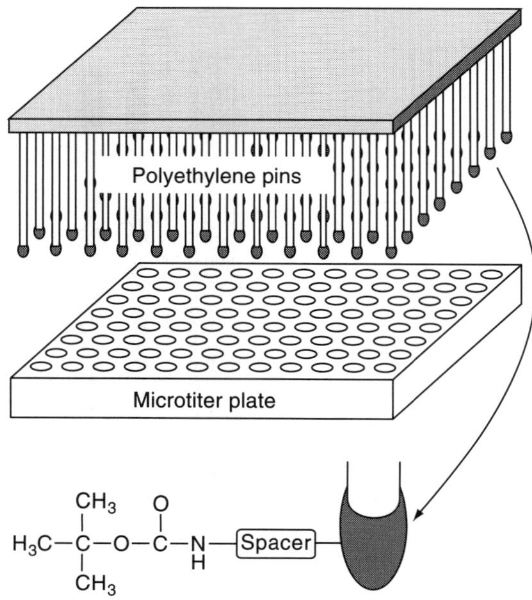

Figure 2.21 Multiple peptide synthesis on pins. The location of the polyethylene pins is complementary to the wells of a microtiter plate. The synthesis takes place on the functionalized tips of the pins by dipping them into the wells that are filled with the reaction solutions or solvents, respectively.

The significance of the pin method lies less in the study of the structure–activity relationships of hormone analogs, but rather in the immunoanalytical area, due to these limitations. Antibodies, for instance, can be tested to determine certain antigenic areas on protein surfaces. For this, overlapping tetra- to nonapeptide sequences of the protein to be investigated, are synthesized on the pins. Afterwards, mono- or polyclonal antisera are distributed in ELISA plates and incubated with the pin block. The antibodies in the antisera bind preferably to those pins that represent an antigenic protein subsequence (i.e. a so-called continuous epitope). One can consequently map the entire protein with regard to its antigenicity because the synthetic peptides extend over the entire protein. In the case of monoclonal antibodies, the continuous binding region can be determined by this, right down to the exact amino acids (Figure 2.22). This special type of ELISA is frequently referred to as 'Pepscan'.

To obtain free peptides with the pin method, either a labile anchor has to be coupled to the polyethylene pin before the first amino acid or special enzymatic or acidolytic cleavage sites have to be introduced *C*-terminally [Maeji *et al.*, 1990].

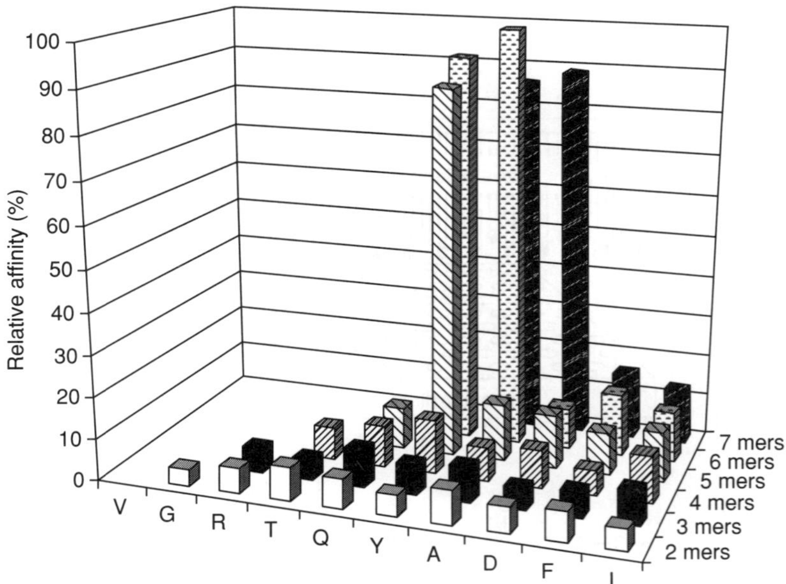

Figure 2.22 Mapping of a continuous epitope. A monoclonal antibody recognizes a certain sequence area of a protein (x-axis). In order to determine the minimal sequence that is recognized, one can synthesize overlapping sequence areas of different lengths by means of a pin synthesis and test the affinity of the antibody for them. Every column represents a certain peptide with the *C*-terminal amino acid indicated on the x-axis and the length indicated on the y-axis. The affinities of the antibodies for the peptides are indicated on the z-axis and permit to conclude that it recognizes the subsequence GRTQY.

2.2.2 TEA BAG SYNTHESIS

In 1985, Houghten introduced the so-called tea bag synthesis of peptides, which allowed the production of relatively large amounts (up to 50 mg) of many different peptides (up to 150) (Figure 2.23) [Houghten, 1985].

At the start of the procedure, the polymeric support is separated by sealing it in a polypropylene mesh – this is where the term tea bag synthesis comes from. One bag is required for each peptide; the bag is labeled with solvent-resistant ink for better differentiation. Polystyrene cross-linked with 1 % DVB is typically used as the resin. In principle, other materials could also be used, if the smallest particle size is not less than the pore width of the mesh. The cleavage of the α-amino protecting group and washing are done for all of the tea bags together in a polypropylene, screw-cap bottle that is adapted to the number of peptides and the washing volume. The bags are sorted according to the amino acid to be coupled, and treated in parallel in separate reaction vessels with the appropriate amino acids and coupling reagents. Two washing steps are performed to remove excess reagents; the bags can be joined together again after

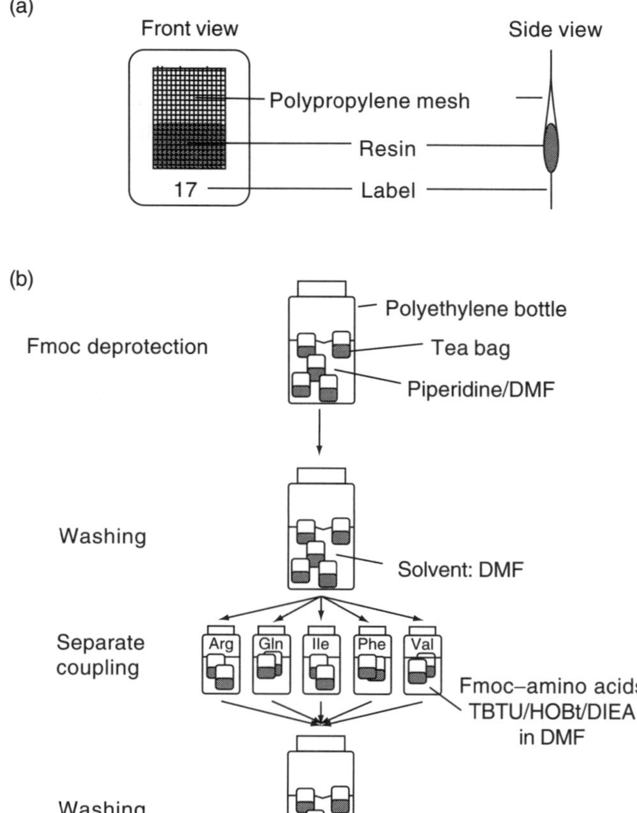

Figure 2.23 Tea bag synthesis. (a) Schematic structure of a tea bag. (b) Synthesis steps for multiple extension of peptides by one amino acid each (following Jung and Beck-Sickinger [1992]).

this. To ease the diffusion in the resin, both the washing and the coupling are done with heavy shaking.

Houghten originally developed the method for the Boc/benzyl strategy with diisopropylcarbodiimide (DIC) activation. In principle, however, the method can be used independently of the protecting group strategy and activation method [Beck-Sickinger *et al.*, 1991].

Based on this variability, the process is especially suited to methodical investigations, because as well as the amino acid activation, the solvents, the coupling schemes and other synthesis parameters can be varied for certain peptides. In addition, the hormone and inhibitor research brought entirely new inspiration, because the method allowed the fast and systematic exchange of each individual amino acid of an active peptide, in order to identify side chains relevant for its action (Figure 2.24). To the same extent, the broadly implemented variation of the chain lengths, as well as the modification of the *N*- and *C*-termini, was facilitated.

The product amount of 30–50 mg raw peptide having a length of more than 10 amino acids is very large in comparison to other parallel synthesis methods. This allows detailed physicochemical analysis including nuclear magnetic resonance (NMR) studies and provides sufficient compound for a extensive biological screening. Further, the setup costs are very low, because no devices are required for the tea bag synthesis.

A major disadvantage of the method, in contrast, lies in the labor intensity, because all of the steps (washing, sorting of the bags, etc) have to be carried out manually. The work can be simplified, however, through computer-supported calculation of the synthesis cycles – especially for the sorting – and automation of the washing steps [Beck-Sickinger *et al.*, 1991]. In addition to the labor intensity, the reaction control creates problems, because it is not practically feasible for a larger number of peptides.

2.2.3 CELLULOSE AS SOLID PHASE

The multiple synthesis of peptides on cellulose discs was described in 1988 by Frank *et al.* [Blankenmeyer-Menge and Frank, 1988; Frank and Döring, 1988],

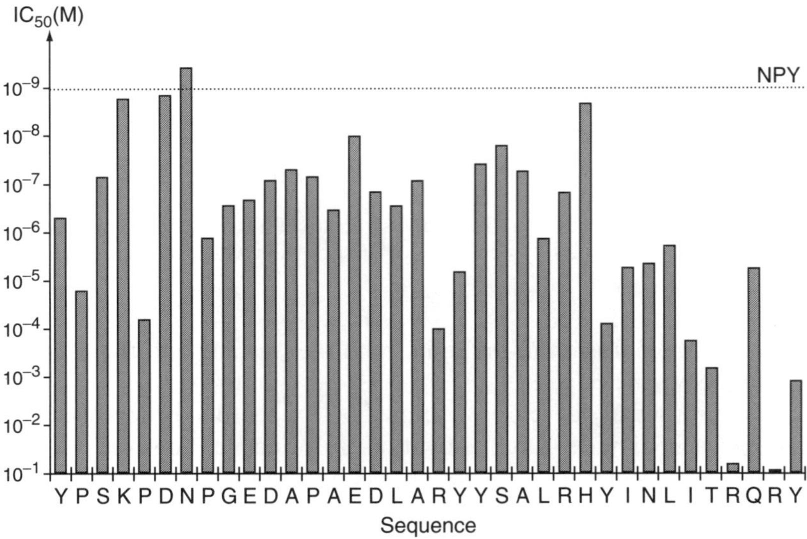

Figure 2.24 Identification of amino acids of neuropeptide Y (NPY) that are important for the binding to its Y_1-receptor, with the aid of a so-called alanine scan [Beck-Sickinger *et al.*, 1994]. Every bar represents the binding affinity of a NPY analog for which the native amino acid in each case was exchanged for alanine. Alanines that are present in the original sequence were replaced by glycines. It can be noted that the *C*-terminal amino acids, in particular Arg^{33} and Arg^{35}, are essential for the interaction of NPY with its Y_1-receptor.

but they had already used the material in 1983, even before Geysen and Houghten, for the simultaneous synthesis of oligonucleotides [Frank *et al.*, 1983]. For preparation of the synthesis the free hydroxyl groups of the cellulose were modified with TFA-labile *p*-alkoxybenzyl ester anchors. Discs were punched out of the paper and stacked in columns. The synthesis of the peptides then took place in a continuous flow synthesizer by means of the Fmoc/*t*Bu strategy. Up to 100 columns can be hooked up in parallel and one peptide is synthesized in each case per column. This method is particularly suited for systematic sequence variations; a computer program supplies the correct combination for a minimal number of cycles.

A different variant of the multiple synthesis of peptides on cellulose was introduced by Eichler *et al.*, in which they used paper strips for which the hydroxyl groups were esterified by *N*-terminally protected amino acid chlorides [Eichler *et al.*, 1989]. The synthesis of the peptides takes place in analogy to the tea bag strategy. Both the use of the Boc/benzyl and the Fmoc/*t*Bu strategy is possible. Subsequently, the synthesized peptides can either be cleaved from the paper or remain on the paper for antibody binding tests with ELISAs after the removal of the side chain protecting groups (dot blot).

The advantages of this method in comparison to the tea bag strategy particularly lie in the simpler handling – cutting of the paper instead of filling and sealing the bags – as well as lower costs due to the use of less expensive materials and lower consumption of solvents. The low mechanical stability and the polyfunctionality of the support have to be accepted as a compromise for this.

2.2.4 AUTOMATION OF THE SOLID PHASE SYNTHESIS OF PEPTIDES

One of the main reasons for the success of the solid phase peptide synthesis, and thereby also the simultaneous multiple peptide synthesis and – as we will see later on – the synthesis of peptide libraries consists in the capability for automation of the process. This is based on the fact that the deprotection, washing and coupling steps recur in each synthesis cycle and can be carried out by robots in principle. Already in 1966, Merrifield developed a device that allowed the semi-automatic synthesis of individual polypeptides [Merrifield *et al.*, 1966]. The automation of the simultaneous multiple peptide synthesis could be realized soon after the introduction of the multiple approaches [Schnorrenberg and Gerhardt, 1989; Gausepohl *et al.*, 1990] (see Chapter 5).

2.3 PEPTIDE LIBRARIES

An approach that went much further was pursued based on the development of the simultaneous multiple peptide synthesis: the synthesis of so-called peptide libraries. In an initial step, a multitude of diverse peptides has to be synthesized. At this stage, however, it is not a matter of exactly knowing the identity of each individual compound which is in contrast to the multiple syntheses. Thereafter follows the screening of these peptides, e.g. the investigation of their affinity for an antibody or their receptor activity. In the last step, the peptide with the 'most interesting' properties has to be identified, i.e. its exact sequence has to be determined.

In the meanwhile, an entire series of methods exists for the realization of the peptide library concept. A classification into three categories seems to be useful:

- libraries that are based on the use of mixtures;
- parallel libraries (arrays);
- biological libraries (see Chapter 6).

The overview in Table 2.4 summarizes the fundamental publications with regard to peptide library strategies.

Table 2.4 Overview of basic publications for the synthesis of peptide libraries.

Author (year)	Name	Polymer	Type	Reference
Geysen (1986)	Pin synthesis	Functionalized polyethylene pins	Mixture library (amino acids)	[Geysen *et al.*, 1986]
Furka (1988)	Portioning–mixing	Polystyrene–divinylbenzene	Mixture library (peptides on polymer)	[Furka *et al.*, 1988a, b; 1991]
Scott Cwirla Devlin (1990)	Phage display	—	Biological library	[Cwirla *et al.*, 1990; Devlin *et al.*, 1990; Scott and Smith, 1990]
Houghten (1991)	Divide-couple-recombine	Polystyrene–divinylbenzene in tea bags	Mixture library (peptides on polymer)	[Houghten *et al.*, 1991]
Lam (1991)	One bead one peptide (split synthesis)	Polystyrene–divinylbenzene beads	Mixture library (peptides on polymer)	[Lam *et al.*, 1991]
Fodor (1991)	Light-directed spatially addressable parallel chemical synthesis	Functionalized glass plates	Parallel library	[Fodor *et al.*, 1991]
Frank (1992)	Spot synthesis	Cellulose	Parallel library	[Frank, 1992]

2.3.1 PEPTIDE LIBRARIES THAT ARE BASED ON THE USE OF MIXTURES

The first library of this type was realized by Geysen *et al.* in 1986 [Geysen *et al.*, 1986]. The objective of the approach was to identify a discontinuous viral epitope of a monoclonal antibody. For this it is important to know that an epitope can also be mimicked by peptides that are not part of the original protein sequence. Antibody-binding peptides of this type are often referred to as ‘mimotopes’.

Instead of individual peptides – as in 1984 – Geysen *et al.* synthesized pin-bound peptide mixtures that represented millions of different individual peptides. To achieve this, they no longer coupled individual amino acids, but instead mixtures of all 20 naturally occurring amino acids to the pins. By using these amino acid mixtures instead of individual amino acids for coupling at six sequence positions they obtained peptide mixtures that theoretically contained $20^6 = 64\,000\,000$ peptides.

This revolutionary synthesis strategy, which consequently necessitated new concepts for the final identification of an active compound from the mixtures, marked the birth of combinatorial chemistry (see Chapter 4).

A fundamental problem in using amino acid mixtures for creating ‘mixed’ positions is the different coupling efficiency of the individual amino acids. This leads in the end to different concentrations of the possible individual peptides in the mixture or even to the complete lack of some individual sequences. To solve this problem, Geysen *et al.* varied the relative concentration of each individual amino acid in accordance with its coupling efficiency.

In 1988, Furka introduced a more accurate, although more laborious, possibility: the ‘portioning–mixing’ method presented in Figure 2.25 [Furka *et al.*, 1988a, 1988b, 1991]. A mixture of peptides of the same length, but different sequence, was synthesized by means of a normal solid phase peptide synthesis by portioning the polymeric support evenly before the coupling of ‘mixed’ positions. Of course, the number of portions corresponded to the number of amino acids that were to be varied at each position. One of the amino acids was then selectively coupled in each vessel. Thus, identical coupling yields were achieved, because the competition between fast and slow coupling reactions was eliminated. After this, the polymer was mixed again, the resulting polymer-bound mixture was washed, and the *N*-terminal protecting group was removed. After reportioning, the procedure was repeated until the desired length of the peptides was achieved (see Chapter 4).

To better exhaust the potential of Furka’s idea for the screening of peptide libraries Lam *et al.* developed the so-called ‘one bead one peptide’ concept in 1991. The synthesis principle was more or less identical to that of Furka, but since the practical application was different it was called ‘split synthesis’ (Figure 2.26) [Lam *et al.*, 1991]. The polymer is used in the form of small granules (so-called beads, Ø ≈ 100–200 μm) that are portioned before every coupling step.

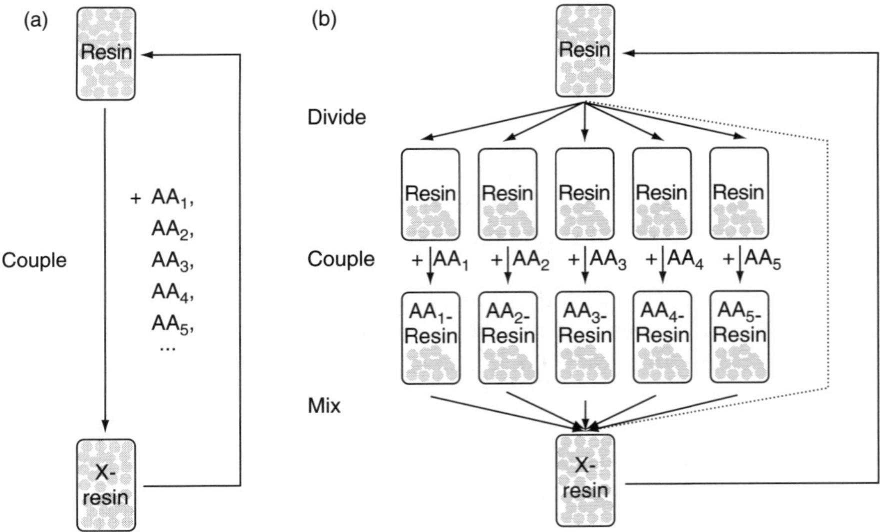

Figure 2.25 Methods for producing so-called 'mixed' positions. (a) Coupling of amino acid mixtures. (b) Portioning–mixing strategy according to Furka *et al.* [1988a,b; 1991]: portioning of the resin, separate coupling of individual amino acids, mixing of the amino acid resins. The different coupling tendency of the individual amino acids can lead in the case of method (a) to different concentrations of the possible individual peptides in the mixture. Method (b) circumvents this problem by separate coupling of the amino acids, but it is more laborious.

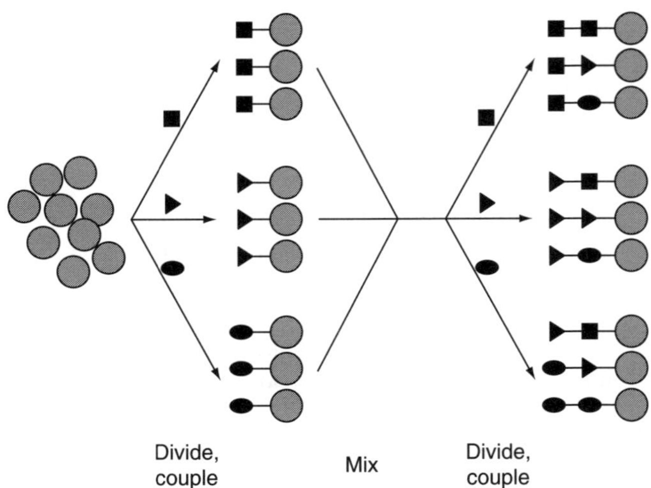

Figure 2.26 'One bead one peptide' strategy. Small polymer granules, so-called beads, serve as polymeric supports. Before every coupling step, they are portioned, the amino acids are separately coupled and recombined after this. As a result, one bead exclusively carries peptides of the same sequence, that varies from bead to bead.

After separate coupling of the amino acids, the beads are recombined and only portioned again to couple the next amino acid. As a result, every individual bead only contains peptides of the same sequence, which varies from bead to bead (see Chapter 4).

At the same time as Lam *et al.*, Houghten *et al.* suggested a further method for building up a peptide library and for identifying a certain biologically active sequence that was based on the tea bag synthesis developed by them (see Section 2.2.2). In analogy to Furka's idea, they produced a peptide library that provided relatively large amounts of free peptides for the testing, which is in contrast to the methods of Geysen and Lam. To distinguish their strategy from the other approaches they named it 'divide, couple, and recombine' (DCR) [Houghten *et al.*, 1991] (see Chapter 4).

2.3.2 PARALLEL LIBRARIES (ARRAYS)

There is more or less no clear-cut border between the transition from the simultaneous multiple peptide synthesis to the parallel peptide library – also called 'array' in new terminology. Thus, Geysen's parallel synthesis of defined peptides on polyethylene pins in 1984 can be viewed as the first peptide library ever, or of this type (see Section 2.1 of this chapter). Fodor *et al.* in 1991 introduced the first process that had, without question, a parallel peptide library as its objective (Figure 2.27) [Fodor *et al.*, 1991]. This completely new approach combined the multiple solid phase peptide synthesis with elements of photolithography and photochemistry and became known as the 'light-directed spatially addressable parallel chemical synthesis'. The solid phase synthesis is carried out on functionalized glass plates in combination with a photolabile α-amino protecting group. Very small areas can now be irradiated with light of a suitable wavelength in a directed way using appropriate photolithographic masks. The unmasked α-amino protecting groups become cleaved. In a subsequent coupling step the chain elongation can only take place in the irradiated areas. Following this routine, up to 40 000 sequences can be synthesized next to each other on a glass plate with an area of 1 cm^2. The final screening can be done, for example, using fluorescently labeled antibodies. Their affinity for a peptide can be proven under the fluorescence microscope.

The principle of a peptide library can be illustrated quite well once again with this method. The sequence data of the 40 000 different peptides are of no interest in the beginning (in principle they would be known in this case, because of the computer-controlled masking). The sequence information only becomes significant after the screening. However, only the data set of the 'most interesting' peptide is needed, which can now be selectively called up from the whole mass of sequence information.

A further technique for the synthesis of parallel peptide libraries or arrays was introduced by Frank: the so-called 'spot synthesis' [Frank, 1992]. The

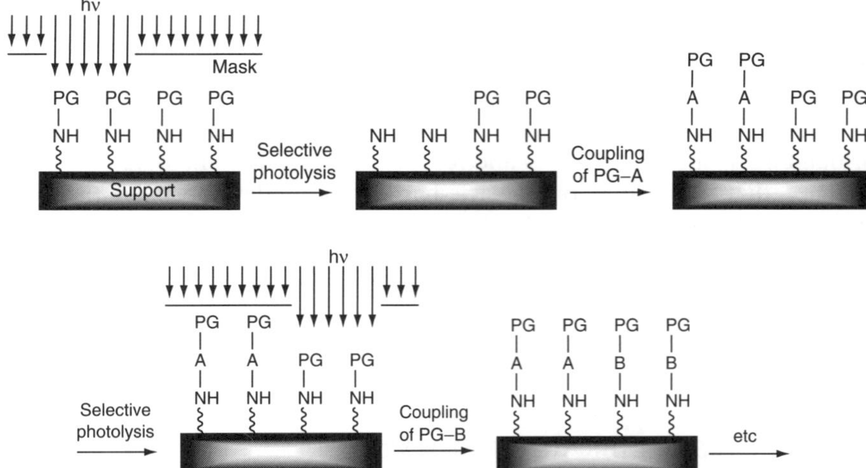

Figure 2.27 Parallel multiple solid phase peptide synthesis using photolithographic techniques (light-directed spatially addressable chemical synthesis). The synthesis is carried out on functionalized glass plates in combination with a photolabile α-amino protecting group (PG). A high number of different sequences can be synthesized next to each other in a very small area by irradiation of the glass plates with light of a suitable wavelength using appropriate masks. Thus, unmasked amino protecting groups are removed. In the subsequent coupling step a chain extension only takes place in the areas that have been irradiated (following Fodor *et al.*, [1991]).

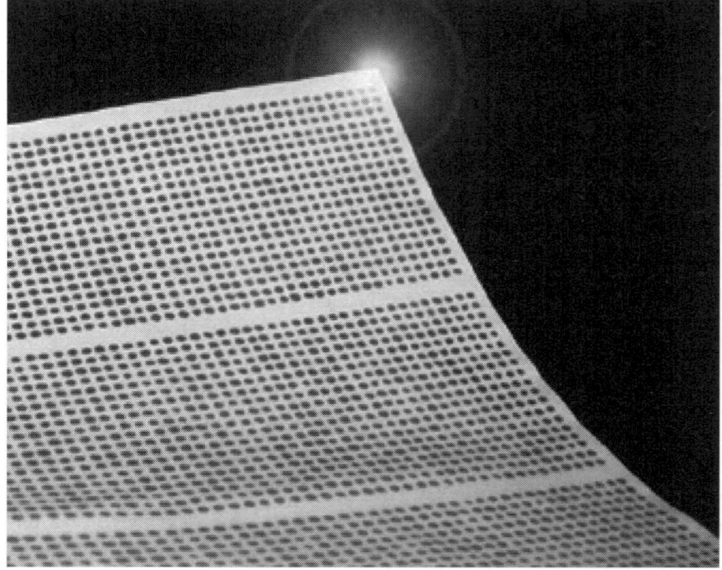

Figure 2.28 Spot synthesis. Membrane with 8000 spots (individual compounds or compound mixtures) on the surface of a DIN A4 sheet (the illustration was kindly made available by Jerini, Berlin, Germany).

peptides are synthesized on amino-functionalized cellulose paper sheets by manual or automatic pipetting of the reagents at exactly-'defined' positions (spots). The washing and deprotection steps are, in contrast, executed by dipping the paper sheets in appropriate solutions. As a result, arrays that consist of hundreds of different polymer-bound peptides can be produced on an area of a few square centimeters (Figure 2.28). The screening of the spots is usually done via antibody tests in combination with radioactive or fluorescence labeling or induction of a color reaction, because the arrays are frequently used for epitope analysis.

REFERENCES

Albericio, F., Kneib-Cordonier, N., Biancalana, S., Gera, L., Masada, R. I., Hudson, D. and Barany, G. (1990). Preparation and application of the 5-(4-(9-fluorenylmethoxycarbonyl)aminomethyl-3,5-dimethoxyphenoxy)valeric acid (PAL) handle for the solid-phase synthesis of C-terminal peptide amides under mild conditions. *J. Org. Chem.* **55**, 3730–3743.

Alsina, J., Chiva, C., Ortiz, M., Rabanal, F., Giralt, E. and Albericio, F. (1997). Active carbonate resins for solid-phase synthesis through the anchoring of a hydroxyl function. Synthesis of cyclic and alcohol peptides, *Tetrahedron Lett.* **38**, 883–886.

Altmann, K.-H. and Mutter, M. (1993). Die chemische Synthese von Peptiden und proteinen, *Chem. unserer Zeit* **27**, 274–286.

Arshady, R., Atherton, E., Clive, D. L. and Sheppard, R. C. (1981). Peptide synthesis. Part 1. Preparation and use of polar supports based on poly(dimethylacrylamide), *J. Chem. Soc., Perkin Trans.* **1**, 529–537.

Atherton, E., Brown, E., Sheppard, R. C. and Rosevear, A. (1981). A physically supported gel polymer for low-pressure, continuous flow solid phase reactions. Application to solid phase peptide synthesis, *J. Chem. Soc. Chem. Commun.* 1151–1152.

Atherton, E., Clive, D. L. and Sheppard, R. C. (1975). Polyamide supports for polypeptide synthesis, *J. Am. Chem. Soc.* **97**, 6584–6585.

Backes, B. J. and Ellman, J. A. (1999). An alkanesulfonamide 'safety-catch' linker for solid-phase synthesis, *J. Org. Chem.* **64**, 2322–2330.

Backes, B. J., Virgilio, A. A. and Ellman, J. A. (1996). Activation method to prepare a highly reactive acylsulfonamide 'safety-catch' linker for solid-phase synthesis, *J. Am. Chem. Soc.* **118**, 3055–3056.

Bannwarth, W., Huebscher, J. and Barner, R. (1996). A new linker for primary amines applicable to combinatorial approaches, *Bioorg. Med. Chem. Lett.* **6**, 1525–1528.

Barlos, K., Gatos, D., Kallitsis, J., Papaphotiu, G., Sotiriu, P., Wenqing, Y. and Schaefer, W. (1989a). Synthesis of protected peptide-fragments using substituted triphenylmethyl resins, *Tetrahedron Lett.* **30**, 3943–3946.

Barlos, K., Gatos, D., Kapolos, S., Papaphotiu, G., Schaefer, W. and Wenqing, Y. (1989b). Esterification of partially protected peptide-fragments with resins – utilization of 2-chlorotritylchloride for synthesis of Leu-15-Gastrin-1, *Tetrahedron Lett.* **30**, 3947–3950.

Bayer, E. and Rapp, W. (1986). New polymer supports for solid–liquid-phase peptide synthesis. In: Voelter, W., Bayer, E., Ovchinnikov, Y. A. and Ivanov, V. T. (eds), *Chemistry of Peptides and Proteins*, vol. 3, Walter de Gruyter, Berlin, pp. 3–8.

Beck-Sickinger, A. G., Duerr, H. and Jung, G. (1991). Semiautomated T-bag peptide synthesis using 9-fluorenyl-methoxycarbonyl strategy and benzotriazol-1-yl-tetramethyl-uronium tetrafluoroborate activation, *Pept. Res.* **4**, 88–94.

Beck-Sickinger, A. G., Wieland, H. A., Wittneben, H., Willim, K. D., Rudolf, K. and Jung, G. (1994). Complete L-alanine scan of neuropeptide Y reveals ligands binding to Y1 and Y2 receptors with distinguished conformations, *Eur. J. Biochem.* **225**, 947–958.

Berg, R. H., Almdal, K., Batsberg Pedersen, W., Holm, A., Tam, J. P. and Merrifield, R. B. (1989). Long-chain polystyrene-grafted polyethylene film matrix: a new support for solid-phase peptide synthesis, *J. Am. Chem. Soc.* **111**, 8024–8026.

Bernatowicz, M. S., Daniels, S. B. and Köster, H. (1989). A comparison of acid labile linkage agents for the synthesis of peptide *C*-terminal amides, *Tetrahedron Lett.* **30**, 4645–4648.

Blackburn, C. (2000). Solid supports for the synthesis of peptides and small molecules, 197–273. In: *Solid-Phase Synthesis – A Practical Guide* (Edited by Kates, S. A. and Albericio, F.), Marcel Dekker, New York.

Blankenmeyer-Menge, B. and Frank, R. (1988). Simultaneous multiple synthesis of protected peptide fragments on 'allyl'-functionalized cellulose disc supports, *Tetrahedron Lett.* **29**, 5871–5874.

Brase, S. and Dahmen, S. (2000). Traceless linkers – only disappearing links in solid-phase organic synthesis? *Chem. Eur. J.* **6**, 1899–1905.

Bycroft, B. W., Chan, W. C., Chhabra, S. R. and Hone, N. D. (1993). A novel lysine protecting procedure for continuous flow solid phase synthesis of branched peptides, *J. Chem. Soc. Chem. Commun.* 778–779.

Chenera, B., Finkelstein, J. A. and Veber, D. F. (1995). Protodetachable arylsilane polymer linkages for use in solid-phase organic-synthesis, *J. Am. Chem. Soc.* **117**, 11 999–12 000.

Cwirla, S. E., Peters, E. A., Barrett, R. W. and Dower, W. J. (1990). Peptides on phage: a vast library of peptides for identifying ligands, *Proc. Natl. Acad. Sci. USA* **87**, 6378–6382.

DeGrado, W. F. and Kaiser, E. T. (1980). Polymer-bound oxime esters as supports for solid-phase peptide synthesis. Preparation of protected peptide fragments, *J. Org. Chem.* **45**, 1295–1300.

Devlin, J. J., Panganiban, L. C. and Devlin, P. E. (1990). Random peptide libraries: a source of specific protein binding molecules, *Science* **249**, 404–406.

Eichler, J., Beyermann, M., Bienert, M. and Lebl, M. (1989). Simultaneous peptide synthesis using cellulose paper as support material, 205–207. In: *Peptides 1988 (Proceedings of the 20th European Peptide Symposium)* (Edited by G. Jung and E. Bayer), Walter de Gruyter, Berlin.

Fodor, S. P., Read, J. L., Pirrung, M. C., Stryer, L., Lu, A. T. and Solas, D. (1991). Light-directed, spatially addressable parallel chemical synthesis, *Science.* **251**, 767–773.

Frank, R. (1992). Spot-synthesis an easy technique for the positionally addressable parallel chemical synthesis on a membrane support, *Tetrahedron* **48**, 9217–9232.

Frank, R. and Döring, R. (1988). Simultaneous multiple peptide synthesis under continuous flow conditions on cellulose paper discs as segmental solid supports, *Tetrahedron* **44**, 6031–6040.

Frank, R., Heikens, W., Heisterberg-Moutsis, G. and Bloecker, H. (1983). A new general approach for the simultaneous chemical synthesis of large numbers of oligonucleotides: segmental solid supports., *Nucleic. Acids Res.* **11**, 4365–4377.

Furka, Á., Sebestyén, F., Asgedom, M. and Dibó, G. (1988a). *14th Int. Congr. Biochem.* Prague, Czechoslovakia, p. 47.

Furka, Á., Sebestyén, F., Asgedom, M. and Dibó, G. (1988b). *10th Int. Symp. Med. Chem.* Budapest, Hungary, p. 168.

Furka, Á., Sebestyén, F., Asgedom, M. and Dibó, G. (1991). General method for rapid synthesis of multicomponent peptide mixtures, *Int. J. Pept. Protein Res.* **37**, 487–493.

Gausepohl, H., Kraft, M., Boulin, C. and Frank, R. W. (1990). A robotic workstation for automated multiple peptide synthesis, 487–490. In: Epton, R. (ed.), *Innovation and Perspectives in Solid Phase Synthesis: Peptides, Polypeptides and Oligonucleotides, Macro-Organic Reagents*, SPCC, Birmingham, UK.

Geysen, H. M., Meloen, R. H. and Barteling, S. J. (1984). Use of peptide synthesis to probe viral antigens for epitopes to a resolution of a single amino acid, *Proc. Natl. Acad. Sci. USA* **81**, 3998–4002.

Geysen, H. M., Rodda, S. J. and Mason, T. J. (1986). *A priori* delineation of a peptide which mimics a discontinuous antigenic determinant, *Mol. Immunol.* **23**, 709–715.

Guillier, F., Orain, D. and Bradley, M. (2000). Linkers and cleavage strategies in solid-phase organic synthesis and combinatorial chemistry, *Chem. Rev.* **100**, 2091–2157.

Hoekstra, W. J., Greco, M. N., Yabut, S. C., Hulshizer, B. L. and Maryanoff, B. E. (1997). Solid-phase synthesis via *N*-terminal attachment to the 2-chlorotrityl resin, *Tetrahedron Lett.* **38**, 2629–2632.

Holmes, C. P. (1997). Model studies for new *o*-nitrobenzyl photolabile linkers: substituent effects on the rates of photochemical cleavage, *J. Org. Chem.* **62**, 2370–2380.

Holmes, C. P. and Jones, D. G. (1995). Reagents for combinatorial organic synthesis: development of a new *o*-nitrobenzyl photolabile linker for solid phase synthesis, *J. Org. Chem.* **60**, 2318–2319.

Houghten, R. A. (1985). General method for the rapid solid-phase synthesis of large numbers of peptides: specificity of antigen–antibody interaction at the level of individual amino acids, *Proc. Natl. Acad. Sci. USA* **82**, 5131–5135.

Houghten, R. A., Pinilla, C., Blondelle, S. E., Appel, J. R., Dooley, C. T. and Cuervo, J. H. (1991). Generation and use of synthetic peptide combinatorial libraries for basic research and drug discovery, *Nature* **354**, 84–86.

Ingenito, R., Bianchi, E., Fattori, D. and Pessi, A. (1999). Solid phase synthesis of peptide *C*-terminal thioesters by Fmoc/*t*-Bu chemistry, *J. Am. Chem. Soc.* **121**, 11 369–11 374.

James, I. W. (1999). Linkers for solid phase organic synthesis, *Tetrahedron* **55**, 4855–4946.

Jung, G. and Beck-Sickinger, A. G. (1992). Multiple peptide-synthesis methods and their applications, *Angew. Chem. Int. Ed. Engl.* **31**, 367–383.

Kenner, G. W., McDermott, J. R. and Sheppard, R. C. (1971). The safety catch principle in solid-phase synthesis, *J. Chem. Soc. Chem. Commun.* 636–637.

Krchnák, V., Vagner, J. and Mach, O. (1989). Multiple continuous-flow solid-phase peptide synthesis. Synthesis of an HIV antigenic peptide and its omission analogues, *Int. J. Pept. Protein Res.* **33**, 209–213.

Kunz, H. and Dombo, B. (1988). Solid-phase synthesis of peptides and glycopeptides on polymeric supports with allylic anchor groups, *Angew. Chem. Int. Ed. Engl.* **27**, 711–713.

Kurth, M. J., Randall, L. A. A., Chen, C. X., Melander, C., Miller, R. B., McAlister, K., Reitz, G., Kang, R., Nakatsu, T. and Green, C. (1994). Library-based lead compound discovery – antioxidants by an analogous synthesis deconvolutive assay strategy, *J. Org. Chem.* **59**, 5862–5864.

Lam, K. S., Salmon, S. E., Hersh, E. M., Hruby, V. J., Kazmierski, W. M. and Knapp, R. J. (1991). A new type of synthetic peptide library for identifying ligand-binding activity, *Nature* **354**, 82–84.

Lebl, M. and Eichler, J. (1989). Simulation of continuous solid phase synthesis: synthesis of methionine enkephalin and its analogs, *Pept. Res.* **2**, 297–300.

Letsinger, R. L., Kornet, M. J., Mahadevan, V. and Jerina, D. M. (1964). Reactions on polymer supports, *J. Am. Chem. Soc.* **86**, 5163–5165.

Maeji, N. J., Bray, A. M. and Geysen, H. M. (1990). Multi-pin peptide synthesis strategy for T cell determinant analysis, *J. Immunol. Methods* **134**, 23–33.

Marsh, I. R., Smith, H. and Bradley, M. (1996). Solid phase polyamine linkers – their utility in synthesis and the preparation of directed libraries against trypanothione reductase, *J. Chem. Soc. Chem. Commun.* 941–942.

Matsueda, G. R. and Stewart, J. M. (1981). A *p*-methylbenzhydrylamine resin for improved solid-phase synthesis of peptide amides, *Peptides* **2**, 45–50.

Meldal, M. (1992). PEGA: a flow stable polyethylene-glycol dimethyl acrylamide copolymer for solid-phase synthesis, *Tetrahedron Lett.* **33**, 3077–3080.

Mergler, M., Tanner, R., Gosteli, J. and Grogg, P. (1988). Peptide synthesis by a combination of solid-phase and solution methods. I: a new very acid-labile anchor group for the solid-phase synthesis of fully protected fragments, *Tetrahedron Lett.* **29**, 4005–4008.

Merrifield, R. B. (1963). Solid phase peptide synthesis. I. The synthesis of a tetrapeptide, *J. Am. Chem. Soc.* **85**, 2149–2154.

Merrifield, R. B. (1985). Solid phase peptide synthesis (Nobel lecture), *Angew. Chem. Int. Ed. Engl.* **24**, 799–810.

Merrifield, R. B., Stewart, J. M. and Jernberg, N. (1966). Instrument for automated synthesis of peptides, *Anal. Chem.* **38**, 1905–1914.

Mitchell, A. R., Kent, S. B. H., Engelhard, M. and Merrifield, R. B. (1978). New synthetic route to *tert*-butyloxycarbonylaminoacyl-4-(oxymethyl)phenylacetamidomethyl-resin, an improved support for solid-phase peptide synthesis, *J. Org. Chem.* **43**, 2845–2852.

Morphy, J. R., Rankovic, Z. and Rees, D. C. (1996). A novel linker strategy for solid-phase synthesis, *Tetrahedron Lett.* **37**, 3209–3212.

Patek, M. and Lebl, M. (1998). Safety-catch and multiply cleavable linkers in solid-phase synthesis, *Biopolymers* **47**, 353–363.

Plunkett, M. J. and Ellman, J. A. (1995). A silicon-based linker for traceless solid-phase synthesis, *J. Org. Chem.* **60**, 6006–6007.

Rademann, J., Groetli, M., Meldal, M. and Bock, K. (1999). SPOCC: resin for solid phase organic chemistry and enzyme reactions, *J. Am. Chem. Soc.* **121**, 5459–5466.

Rapp, W. (1996). PEG grafted polystyrene tentacle polymers: physico-chemical properties and application in chemical synthesis, 425–464. In: *Combinatorial Peptide and Nonpeptide Libraries – a Handbook* (Edited by G. Jung), VCH, Weinheim.

Renil, M. and Meldal, M. (1996). POEPOP and POEPS: inert polyethylene glycol crosslinked polymeric supports for solid phase synthesis, *Tetrahedron Lett.* **37**, 6185–6188.

Rink, H. (1987). Solid-phase synthesis of protected peptide fragments using a trialkoxy-diphenyl-methylester resin, *Tetrahedron Lett.* **28**, 3787–3790.

Schnorrenberg, G. and Gerhardt, H. (1989). Fully automatic simultaneous multiple peptide synthesis in micromolar scale – rapid synthesis of series of peptides for screening in biological assays, *Tetrahedron* **45**, 7759–7764.

Scott, J. K. and Smith, G. P. (1990). Searching for peptide ligands with an epitope library, *Science* **249**, 386–390.

Seitz, O. and Kunz, H. (1997). HYCRON, an allylic anchor for high-efficiency solid phase synthesis of protected peptides and glycopeptides, *J. Org. Chem.* **62**, 813–826.

Sheppard, R. C. and Williams, B. J. (1982). Acid-labile resin linkage agents for use in solid-phase peptide synthesis, *Int. J. Pept. Protein Res.* **20**, 451–454.

Shin, Y., Winan, K. A., Backes, B. J., Kent, S. B. H., Ellman, J. A. and Bertozzi, C. R. (1999). Fmoc-based synthesis of peptide-thioesters: application to the total chemical synthesis of a glycoprotein by native chemical ligation, *J. Am. Chem. Soc.* **121**, 11 684–11 689.

Sieber, P. (1987). A new acid-labile anchor group for the solid-phase synthesis of *C*-terminal peptide amides by the Fmoc method, *Tetrahedron Lett.* **28**, 2107–2110.

Small, P. W. and Sherrington, D. C. (1989). Design and application of a new rigid support for high-efficiency continuous-flow peptide-synthesis, *J. Chem. Soc. Chem. Commun.* 1589–1591.

St Hilaire, P. M. and Meldal, M. (1999). Glycopeptide and oligosaccharide libraries, 291–318. In: *Combinatorial Chemistry – Synthesis, Analysis, Screening* (Edited by G. Jung), Wiley-VCH, Weinheim,

Tam, J. P., Tjoeng, F. S. and Merrifield, R. B. (1979). Multidetachable resin supports for solid-phase fragment synthesis, *Tetrahedron Lett.* **51**, 4935–4938.

Tam, J. P., Tjoeng, F. S. and Merrifield, R. B. (1980). Design and synthesis of multidetachable resin supports for solid phase peptide-synthesis, *J. Am. Chem. Soc.* **102**, 6117–6127.

Thompson, L. A. and Ellman, J. A. (1994). Straightforward and general method for coupling alcohols to solid supports, *Tetrahedron Lett.* **35**, 9333–9336.

Wang, S.-S. (1973). *p*-Alkoxy alcohol resin and *p*-alkoxybenzyloxycarbonylhydrazine resin for solid phase synthesis of protected peptide fragments, *J. Am. Chem. Soc.* **95**, 1328–1333.

Warrass, R. (1999). Solid-phase anchors in organic chemistry, 167–228. In: *Combinatorial Chemistry – Synthesis, Analysis, Screening* (Edited by G. Jung), Wiley-VCH, Weinheim.

Wenschuh, H., Carpino, L. A., Albericio, F., Krause, E., Beyermann, M. and Bienert, M. (1995). Stepwise solid phase synthesis of peptaibols using Fmoc–amino acid fluorides, 287–288. In: *Peptides 1994 (Proceedings of the 23rd European Peptide Symposium)* (Edited by H. L. S. Maia), ESCOM, Leiden, the Netherlands.

Winter, M. (1996). Supports for solid-phase organic synthesis, 465–510. In: *Combinatorial Peptide and Nonpeptide Libraries – a Handbook* (Edited by G. Jung), VCH, Weinheim,

Zaragoza, F. (2000). New sulfur- and selenium-based traceless linkers – more than just linkers?, *Angew. Chem. Int. Ed. Engl.* **39**, 2077–2079.

Zhao, X.-Y., Jung, K. W. and Janda, K. D. (1997). Soluble polymer synthesis: an improved traceless linker methodology for aliphatic C–H bond formation, *Tetrahedron Lett.* **38**, 977–980.

3 Nonpeptide Libraries

3.1 PEPTIDOMIMETIC POLYMER LIBRARIES

One of the main areas for the use of peptide libraries is the identification of new lead compounds in pharmaceutical drug research. However, rapid proteolytic degradation through digestive enzymes, poor resorption and quick excretion present fundamental problems that limit the direct use of bioactive peptides.

One possibility for reducing these problems is the subsequent modification of peptide libraries. Their hydrophilicity can be lowered by permethylation of resin-bound peptides. This improves their pharmacokinetic properties significantly (Figure 3.1). In addition, the diversity of the library is increased. Therefore, this approach is often referred to as the 'libraries from libraries' concept [Ostresh *et al.*, 1994]. The difficult analysis of the products is a disadvantage, because the methylation reactions frequently are incomplete.

A different approach is the synthesis and investigation of so-called biopolymer mimetics. They are based on the structure of peptides, but have a modified backbone and consequently a modified configuration of the side chains as well (Figure 3.2). This results in an increased proteolytic stability and a modified polarity. Thus, the pharmacokinetic properties can be significantly improved.

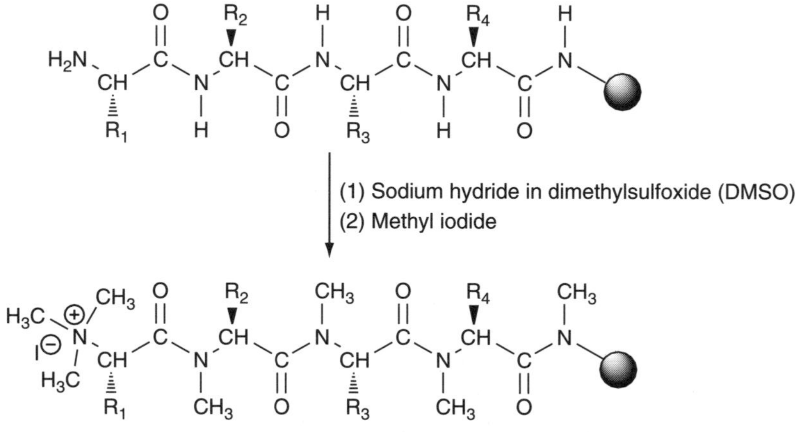

Figure 3.1 Libraries from libraries principle according to Ostresh *et al.*, [1994]. The subsequent modification of a peptide library through permethylation increases the diversity of the library and improves its pharmacokinetic properties.

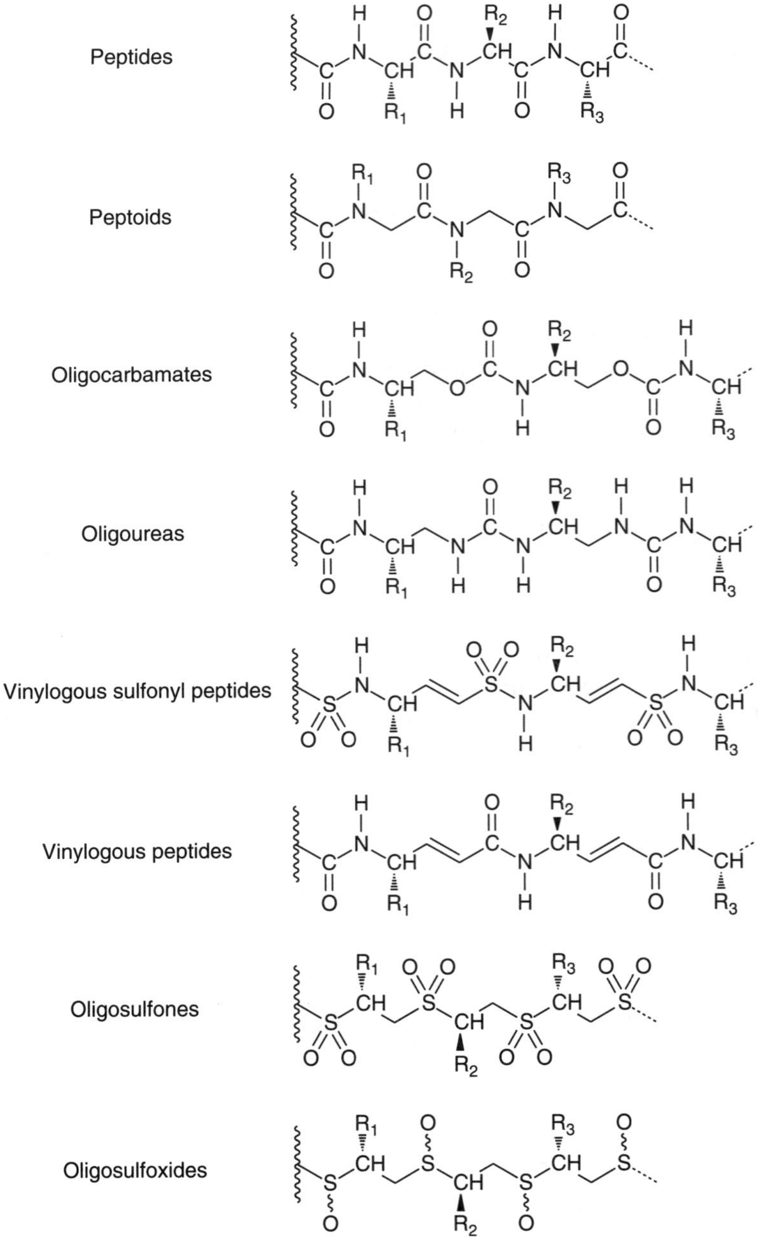

Peptides

Peptoids

Oligocarbamates

Oligoureas

Vinylogous sulfonyl peptides

Vinylogous peptides

Oligosulfones

Oligosulfoxides

Figure 3.2 Overview of the most important representatives of biopolymer mimetics.

3.1.1 PEPTOIDS

Peptoids are *N*-substituted oligoglycines. This means that in contrast to the peptides, the side chains are not connected to the α-carbon, but to the amide nitrogen. As a result, the polymer is achiral and has a high protease resistance [Simon *et al.*, 1992].

The original synthesis of the peptoids is based on 9-fluorenylmethoxycarbonyl (Fmoc)-protected *N*-alkylglycines that have to be synthesized in a laborious way. In the meanwhile, though, a far more elegant method has prevailed that is based on a two-step procedure (Figure 3.3) [Zuckermann *et al.*, 1992]. In the first step, an amide bond is formed by means of α-bromoacetic acid and 1,3-diisopropylcarbodiimide (DIC) activation. The bromine atom is then substituted by a suitable primary amine. A secondary *N*-alkylglycine is formed, which can then react again with α-bromoacetic acid. This so-called 'submonomer' strategy incorporates several advantages:

- no protecting group strategy is required;
- the building blocks are achiral;
- the building blocks are commercially available;
- the reactions have high coupling yields;
- the synthesis can be automated.

Additional diversity can be provided by the subsequent modification of the peptoids, as already described for peptide libraries. Thus, it is possible to obtain isoxazoles or isoxazolines, for example, via a [3 + 2]-cycloaddition of nitrile oxides with alkenyl or alkynyl side chains of peptoids [Pei and Moos, 1994].

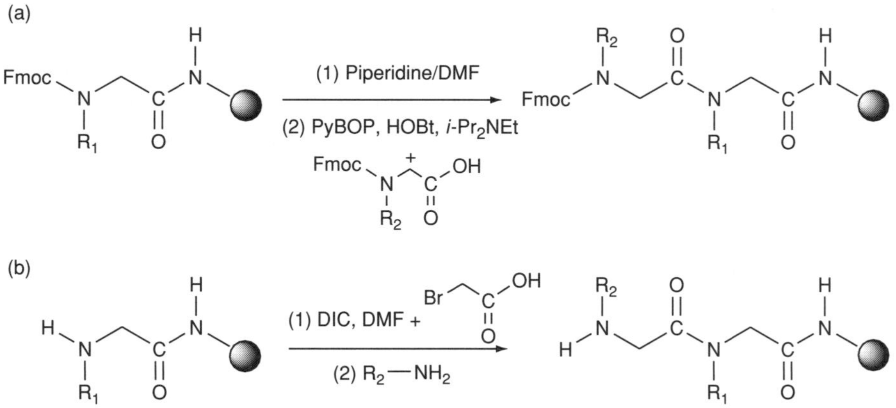

(a)

(1) Piperidine/DMF

(2) PyBOP, HOBt, i-Pr_2NEt

(b)

(1) DIC, DMF +

(2) R_2—NH_2

Figure 3.3 Peptoid synthesis scheme. (a) Classical approach by means of Fmoc-protected *N*-alkylglycines that are expensive to synthesize. (b) Two-step submonomer approach, in which an amide bond is formed by means of α-bromoacetic acid and DIC activation in the first step and, in the second step, the substitution of the bromine by a primary amine.

3.1.2 OLIGOCARBAMATES

The oligocarbamates represent another possibility for creating protease-resistant libraries on a biopolymer basis [Cho *et al.*, 1993]. *N*-protected *p*-nitrophenyl carbonate monomers, which have to be synthesized from the corresponding amino acids, are used as building blocks (Figure 3.4). These monomers can be directly coupled to Rink amide resin by means of 1-hydroxybenzotriazole (HOBt) activation. If Fmoc is used as a protecting group, it can be cleaved with piperidine in analogy to peptide synthesis. The use of the photolabile 6-nitroveratryloxycarbonyl (Nvoc) protecting group is also possible, though. Thus, the library can also be built up as a spatially addressable parallel library using the photolithographic technique already described in Section 2.3.2.

3.1.3 OTHER POLYMERS

Libraries based on other biopolymer mimetics have also been made (Figure 3.2). For synthetic or stability reasons, however, they attracted less attention. Besides oligoureas, vinylogous peptides and vinylogous sulfonyl peptides, this applies to the oligosulfones and oligosulfoxides. More detailed information can be found in the review articles by Thompson and Ellman [1996], Soth and Nowick [1997] and Barron and Zuckermann [1999].

3.2 CARBOHYDRATE LIBRARIES

The natural biopolymers are almost exclusively built up from amino acids, nucleotides or saccharides, but the biochemical role of the oligosaccharides has been investigated far less in comparison to the oligonucleotides and the oligopeptides. This is mainly based on the fact that, up until recently, only the

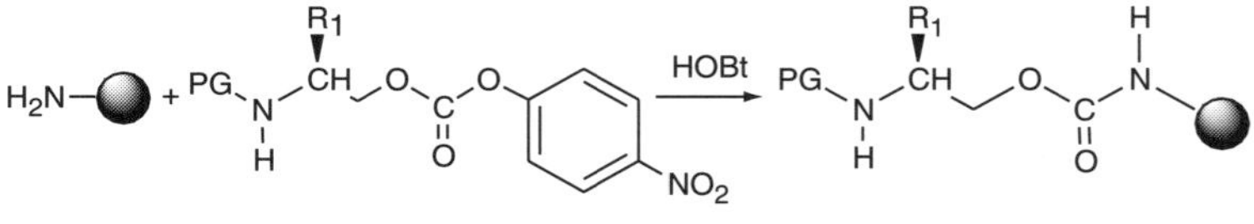

Figure 3.4 Oligocarbamate synthesis scheme. *N*-protected *p*-nitrophenyl carbonate monomers that have to be synthesized from the corresponding amino acids serve as building blocks for the synthesis. The monomers can be directly coupled to the Rink amide resin by means of HOBt activation. If Fmoc is used as the *N*-protecting group (PG), they can be cleaved with piperidine in a fashion analogous to the peptide synthesis. The oligocarbamate synthesis can also take place via a photolithographic process, though, with the use of the photolabile 6-nitroveratryloxycarbonyl (Nvoc) protecting group.

oligonucleotides (information storage and retrieval, i.e. RNA and DNA) and the oligopeptides (catalysis, i.e. enzymes; communication, i.e. hormones and receptors) were accorded a crucial role in biochemical processes. The carbohydrates were merely viewed as energy storages (sugars, starch) or providers of structure (cellulose, chitin). As a result of this, relatively simple synthesis strategies (automated solid phase synthesis or recombinant methods) and sequencing technologies (Edman degradation, Sanger dideoxy method) were developed for oligopeptides or oligonucleotides, which in turn served as the basis for the rapid development of combinatorial libraries of these compound classes.

Since the early 1990s, however, increasingly research results have shown that the carbohydrates also play a definite role in the control of biochemical processes. The proof that an oligosaccharide is the natural ligand of E-selectin, thereby playing a crucial role in cell adhesion, is of central importance in this context [Lowe *et al.*, 1990; Phillips *et al.*, 1990; Walz *et al.*, 1990]. The synthesis of oligosaccharides (Figure 3.5) turned out to be much more difficult, however,

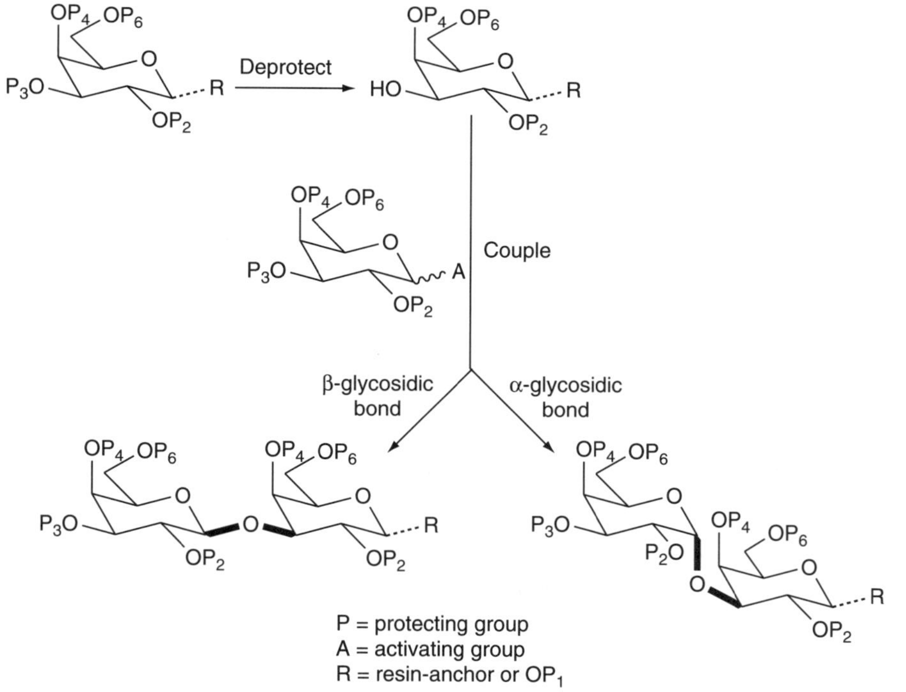

Figure 3.5 The synthesis of oligosaccharides and its problems. In addition to the multitude of reaction centers and difficulties associated with that for the development of a suitable protecting group strategy, the synthesis of pure monosaccharide building blocks also turns out to be difficult. Furthermore, formation of a glycosidic bond is heavily dependent on the respective reaction partners and represents a stereochemical problem, because both the α- and the β-glycosidic bond can arise.

than the synthesis of oligonucleotides and oligopeptides. This is based on various problems:

- The formation of a glycosidic bond is heavily dependent on the reaction partners and usually does not have a quantitative yield.
- The bond formation represents a stereochemical problem, because both the α- and the β-glycosidic bond can be formed.
- Monosaccharides have several hydroxyl groups that can be potential reaction centers, which puts high demands on suitable protecting group strategies.
- The synthesis of pure monosaccharide building blocks turns out to be difficult.

The routine synthesis of carbohydrate libraries is therefore not yet an established practice, although promising efforts are being made worldwide. Kahne's group discovered two *Bauhinia purpurea* lectin ligands, which bind stronger than the natural ligand, by the solid phase synthesis of a disaccharide and trisaccharide library with 1300 members [Liang *et al.*, 1996]. They used a special glycosylation method for this, in which anomeric sulfoxides serve as glycosyl donors (Figure 3.6). Several review articles are recommended as further reading

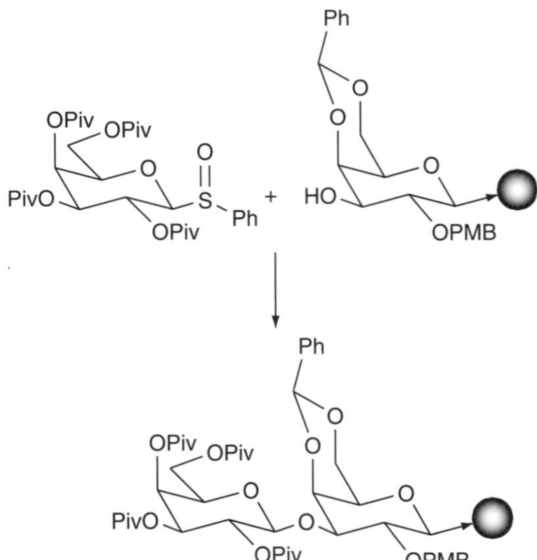

Figure 3.6 Glycosylation with anomeric sulfoxides. The reaction takes place after the addition of 2,6-di-*tert*-butyl-4-methylpyridine in dichloromethane (DCM) at −65 °C and the subsequent addition of trifluoromethyl sulfonic acid (TFMSA) in DCM within 2 hours, while warming to 0 °C. Piv (pivaloyl) and PMB (*p*-methoxybenzyl) are commonly used hydroxyl protecting groups.

matter [Kahne, 1997; Taylor, 1997; Schweizer and Hindsgaul, 1999; Sofia and Silva, 1999; Seeberger and Haase, 2000; St Hilaire and Meldal, 2000].

3.3 SOLID PHASE LIBRARIES OF LOW MOLECULAR WEIGHT COMPOUNDS

Although peptide chemistry has experienced a major upturn since the start of the 1960s because of the use of solid phase techniques, the applications in the area of traditional organic chemistry were limited. A few papers were published on this topic in the 1970s that were, however, not met with a great deal of interest. This was possibly due to the relatively low amounts of product in the solid phase synthesis, for which the analytical methods available at that time were too insensitive. The small amounts also prevented the development of other applications, for example biological screening, because suitable test systems were lacking.

The solid phase organic synthesis only experienced a rebirth in the early 1990s after the establishment of combinatorial libraries based on peptides and oligonucleotides. This could have been due to the biopolymer libraries being given certain limits in their use in pharmaceutical drug research for identifying new lead compounds:

- they only offer relatively limited possibility for diversification;
- they are built up in a linear fashion which leads to a high flexibility;
- they have poor pharmacokinetic properties.

Libraries of low molecular weight compounds suggested themselves as a solution, with which completely new areas of diversity were able to be developed and which in addition have extremely favorable pharmacokinetic properties in part due to their rigidity and small size.

The work of Bunin and Ellman from 1992 concerning the synthesis of 1,4-benzodiazepine derivatives on the solid phase laid the cornerstone of the great deal of importance that is attached to investigation of libraries of low molecular weight compounds [Bunin and Ellman, 1992]. They used a system for this that is based on Geysen's pin synthesis strategy (see Chapter 2, Section 2.1). After the coupling of 2-aminobenzophenones through trifluoroacetic acid (TFA)-labile linkers to the polymer pins, they reacted them with *N*-protected amino acids and alkylating reagents (Figure 3.7). The members of the compound class of 1,4-benzodiazepines are among the most important bioavailable therapeutics (for example Valium®) and show very diverse biological activities. After the establishment and partial optimization of their solid phase synthesis, the researchers constructed pin-bound 1,4-benzodiazepine-2-one libraries that consisted in part of far more than 10 000 individual compounds. The libraries were screened against diverse receptor and enzyme targets in order to generate other potential drug candidates [Bunin *et al.*, 1994, 1996].

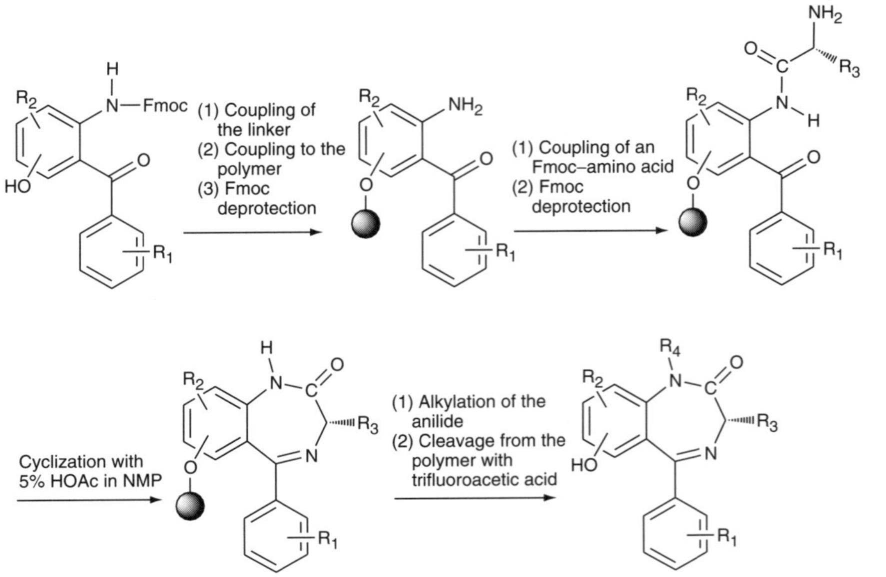

Figure 3.7 Synthesis of a 1,4-benzodiazepine library according to Bunin and Ellman [1992] (NMP = *N*-methylpyrrolidone).

Researchers from Parke-Davis chose a different strategy for the synthesis of 1,4-benzodiazepine derivatives; they started with polymer-bound amino acids (Figure 3.8) [DeWitt *et al.*, 1993]. The so-called diversomer technology was used here for the first time, with which the automated parallel synthesis of organic

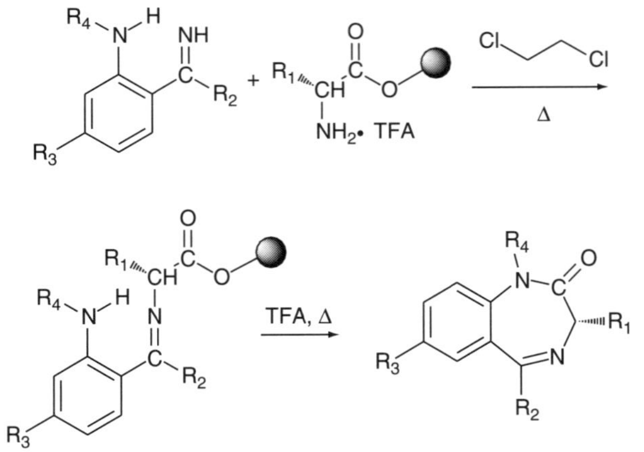

Figure 3.8 Synthesis of a 1,4–benzodiazepine-2-one library according to DeWitt *et al.* [1993]. Imine derivatives of 2–aminobenzophenones are coupled to polymer-bound amino acids. After this, the linear intermediate is cleaved with TFA at 60 °C which leads to the cyclization to 1,4-benzodiazepine-2-ones.

compounds of very diverse compound classes is possible. A little while later, a solid phase library of hydantoins was successfully built up using this technology (Figure 3.9).

The synthesis of libraries of low molecular weight compounds, or the solid phase organic synthesis in general, experienced a rapid upturn since these initial publications. Among others, libraries based on diketopiperazines [Gordon and Steele, 1995], isoquinolinones [Goff and Zuckermann, 1995], 1,4-dihydropyridines [Gordeev *et al.*, 1996], pyrrolidines [Murphy *et al.*, 1995], imidazoles [Sarshar *et al.*, 1996], and phenols [Meyers *et al.*, 1995] are available (Figure 3.10). New compound classes are being added on a nearly daily basis. Two excellent early review articles [Früchtel and Jung, 1996; Thompson and Ellman, 1996] as well as several recent ones [Brown, 1998; Kaldor and Siegel, 1998; Andersson *et al.*, 1999; Felder, 1999; Hall, 1999; Nuss and Renhowe, 1999] are to be pointed out in connection with this.

3.4 SOLUTION PHASE LIBRARY SYNTHESIS OF LOW MOLECULAR WEIGHT COMPOUNDS

Despite the great success of solid phase combinatorial libraries based on simple handling and product isolation, synthesis on polymeric supports involves a fundamental disadvantage: heterogeneous reaction conditions. This can lead to steric hindrance, nonlinear kinetics or even solubility or swelling problems under certain circumstances. In addition, extremely high demands are placed on the compatibility of the individual components: The polymer used, the anchor, the substrates, the protecting group strategy, the activating reagents and the cleaving conditions have to all be coordinated to each other. This led to different approaches for the development of combinatorial libraries in solution.

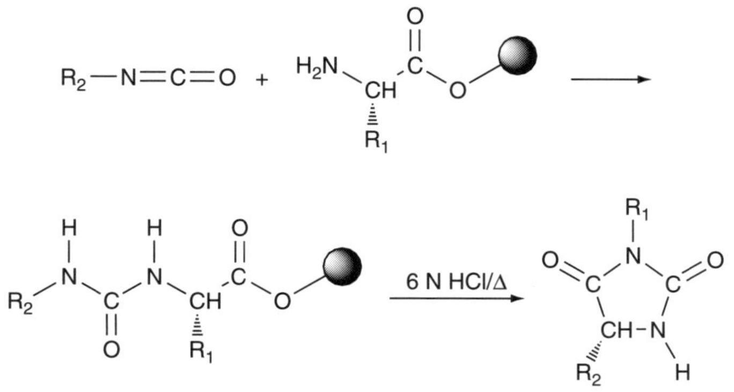

$R_2-N=C=O$ + $H_2N-CH(R_1)-C(=O)-O-$(polymer) ⟶

$R_2-NH-C(=O)-NH-CH(R_1)-C(=O)-O-$(polymer) $\xrightarrow{\text{6 N HCl}/\Delta}$ hydantoin (N–R_1, CH–R_2, N–H)

Figure 3.9 Synthesis of a hydantoin library according to DeWitt *et al.* [1993]. Isocyanates are coupled to polymer-bound amino acids, and the linear intermediates that are formed are cleaved with hydrochloric acid which leads to a cyclization and formation of the hydantoins.

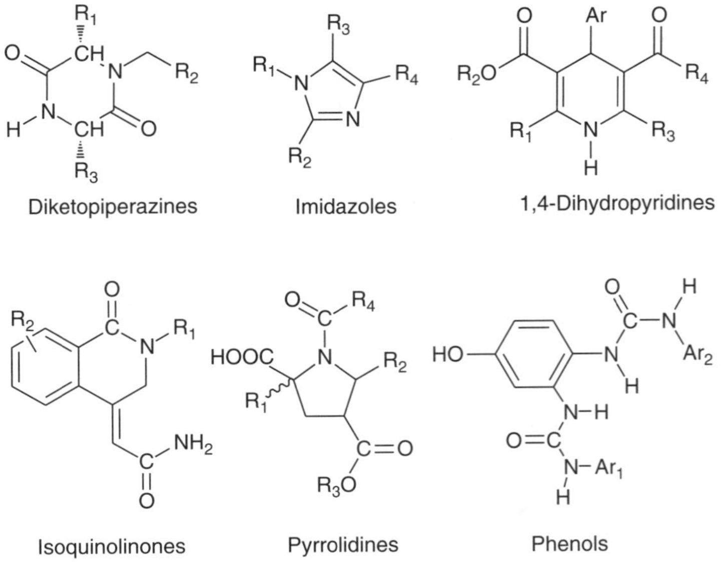

Diketopiperazines Imidazoles 1,4-Dihydropyridines

Isoquinolinones Pyrrolidines Phenols

Figure 3.10 Overview of a few of the most important organic compound classes from which solid phase libraries were synthesized.

3.4.1 SOLUTION PHASE LIBRARY SYNTHESIS OF LOW MOLECULAR WEIGHT COMPOUNDS USING MIXTURES

The first library of this type dates from 1994 and arose through the reaction of 40 individual acid chlorides with an equimolar mixture of 40 nucleophiles (alcohols and amines) (library A) and, in parallel with this, reaction of 40 individual nucleophiles with an equimolar mixture of 40 acid chlorides (library B) [Smith *et al.*, 1994]. Two libraries, A and B, resulted that contained identical $40 \times 40 = 1600$ amides and esters. However, the distribution of these products over the 40 sublibraries containing 40 products each was 'orthogonal' to each other. This means that every possible product exists in each case in exactly one sublibrary of library A (defined acid) and one of library B (defined nucleophile). This approach (the so-called orthogonal deconvolution) represents one possibility for the inevitable decoding or deconvolution of the most active individual compound when using mixtures and is discussed in detail in Section 4.2.3. Independently of this, Pirrung and Chen developed an almost identical approach by synthesizing two carbamate libraries from nine alcohols and six isocyanates, containing $9 \times 6 = 54$ products each. The resulting 'orthogonal' sublibraries were screened for acetylcholinesterase inhibitory activity [Pirrung and Chen, 1995].

An approach based on a different deconvolution strategy (iterative deconvolution, see Section 4.2.3) was developed by Rebek and coworkers [Carell *et al.*, 1994a,b]. Starting with a rigid core molecule containing several reactive functional groups, they created libraries by reacting the core molecules in a single step with mixtures of suitable building blocks (Figure 3.11). A mixture of up to 97 461 different individual compounds arose in this way by treating a xanthene derivative that carried four acid chlorides with a mixture of 21 primary amines.

3.4.2 PARALLEL LIBRARIES IN SOLUTION

Boger and coworkers published one of the first papers on parallel synthesis of a library in solution in 1996 [Cheng *et al.*, 1996]. They obtained up to 1014 discrete end products in this way in a three-step reaction (Figure 3.12). What is remarkable about this method is the very high degree of purity of the end products (more than 90 %) which was achieved through simple liquid–liquid extractions after every reaction step.

Based on a four-component Ugi reaction, Keating and Armstrong developed a concept that was impressive because of an astounding degree of diversity [Keating and Armstrong, 1995, 1996]. Compound classes as diverse as 1,4-benzodiazepine-2,5-diones and monosaccharide derivatives are available here, among other things (Figure 3.13).

More detailed information on the synthesis of arrays by solution phase approaches can be found in the review articles by Ferritto and Seneci [1998], Gayo [1998], Coe and Storer [1999] and Suto [1999].

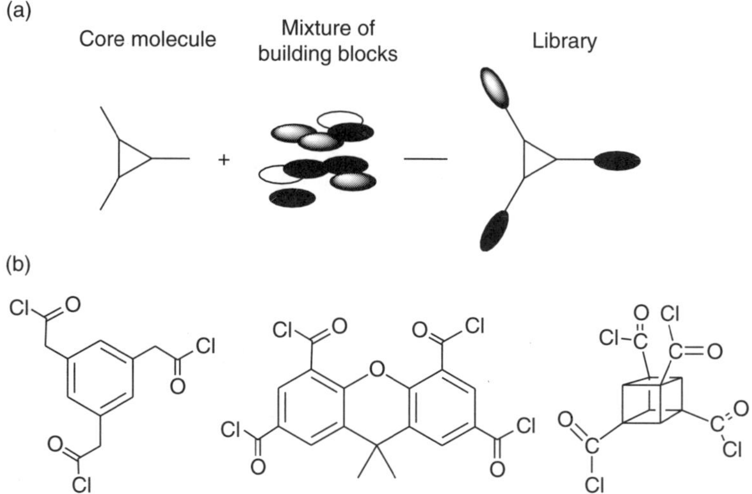

Figure 3.11 Solution phase library synthesis of low molecular weight compounds according to Carrell *et al.* [1994a,b]. (a) Functionalized, rigid core molecules are reacted in a single step with mixtures of suitable building blocks. (b) Examples of core molecules.

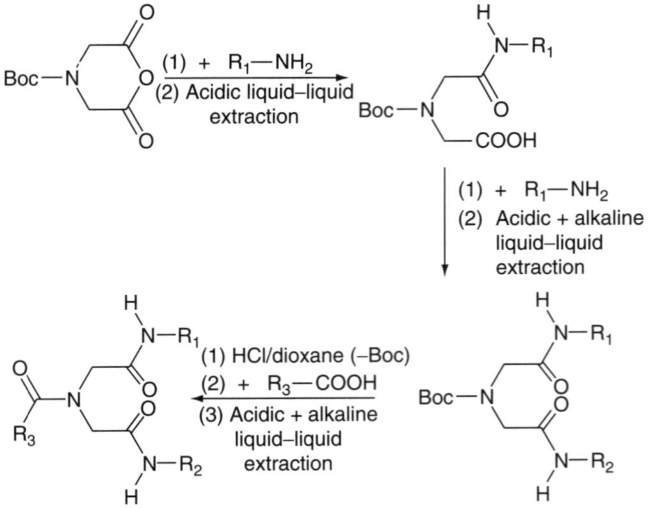

Figure 3.12 Parallel synthesis of a library in solution according to Cheng *et al.* [1996]. A cyclic anhydride that was first converted twice with primary amines and subsequently with carboxylic acids serves as a template.

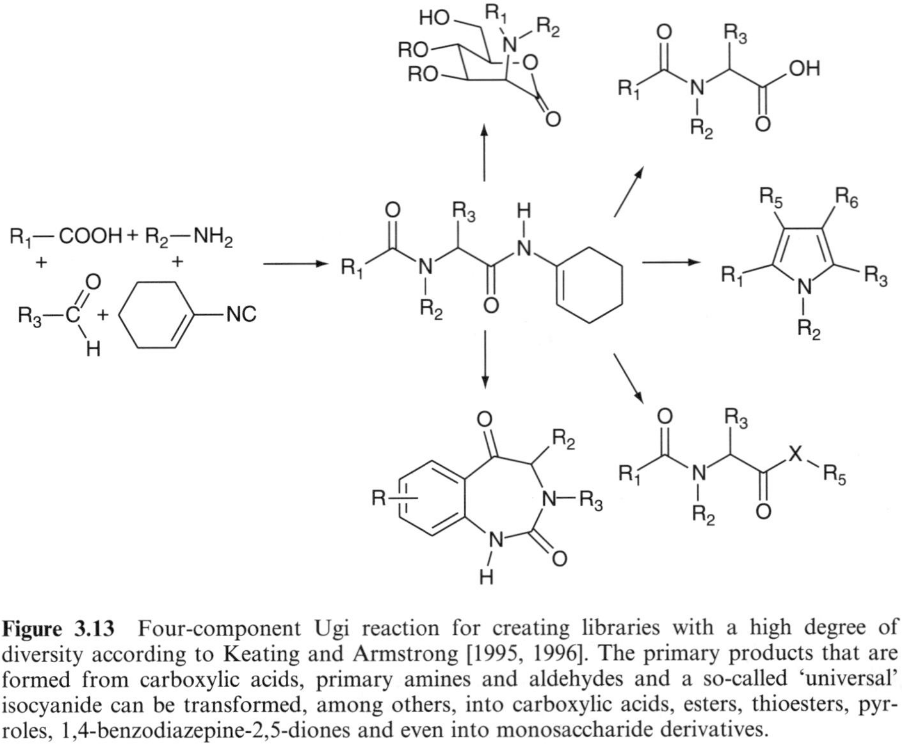

Figure 3.13 Four-component Ugi reaction for creating libraries with a high degree of diversity according to Keating and Armstrong [1995, 1996]. The primary products that are formed from carboxylic acids, primary amines and aldehydes and a so-called 'universal' isocyanide can be transformed, among others, into carboxylic acids, esters, thioesters, pyrroles, 1,4-benzodiazepine-2,5-diones and even into monosaccharide derivatives.

3.4.3 LIBRARY SYNTHESIS IN SOLUTION USING SOLUBLE CARRIER SYSTEMS

The use of soluble polymers (so-called liquid phase synthesis) brings together advantages of the solid phase synthesis – simple sample handling and product isolation – and solution phase synthesis – homogeneous system. After Bayer and Mutter had already done pioneering work in this area in 1972(!) by building up a pentapeptide on polyethylene glycol [Bayer and Mutter, 1972], Janda and coworkers transferred this method to combinatorial chemistry [Han *et al.*, 1995]. They likewise used a polymer based on polyethylene glycol (polyethylene glycol monomethyl ether, $M_r = 5\,kDa$) – a carrier material that combines various advantages:

- soluble in water and many organic solvents;
- acts in a solubilizing fashion;
- relatively inert chemically;
- can be easily separated through size-exclusion chromatography, precipitation with diethyl ether or crystallization.

They built up, on the one hand, a pentapeptide library and, on the other hand, an arylsulfonamide library on the free hydroxyl group (Figure 3.14). Cleaving from the polymer took place with KCN/methanol (peptides) or with 0.5 N NaOH (arylsulfonamides).

Dendrimers can also be used as carrier material. Recently, a small indole library based on the Fischer indole synthesis was able to be synthesized; a Starburst®-polyamidoamine-(PAMAM) dendrimer modified with 4-hydroxymethylbenzoic acid (HMB) was used in the process (Figure 3.15). The dendrimer-bound intermediates were each isolated or purified through size-exclusion chromatography [Kim *et al.*, 1996].

MeO-PEG-OH + O=C=N–C₆H₄–SO₂Cl —(Dibutyltin dilaurate, CH_2Cl_2)→ MeO-PEG-O–C(=O)–NH–C₆H₄–SO₂Cl

↓ + H_2N–R_1, pyridine in CH_2Cl_2

H_2N–C₆H₄–SO₂–NH–R_1 ←(0.5 N NaOH)— MeO-PEG-O–C(=O)–NH–C₆H₄–SO₂–NH–R_1

Figure 3.14 Formation of an arylsulfonamide library with the soluble carrier polyethylene glycol (PEG) according to Han *et al.* [1995]. The separation of the intermediates can take place through size-exclusion chromatography, precipitation with diethyl ether or crystallization.

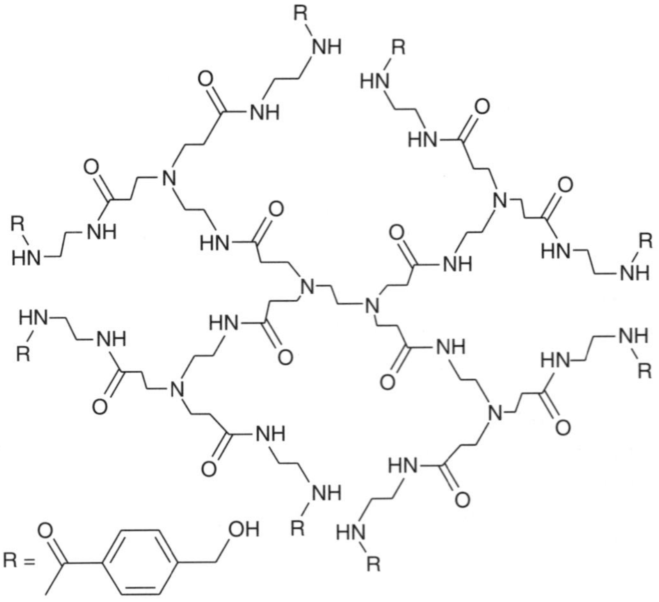

Figure 3.15 Starburst™-polyamidoamine-(PAMAM) dendrimer modified with 4-hydroxymethylbenzoic acid (HMB) anchors for the synthesis of libraries on a soluble carrier.

The building of libraries on so-called fluorous phases is currently a hot topic (Figure 3.16) [Studer *et al.*, 1997a; Curran, 1999]. The potential of this method was shown with the example of a Biginelli reaction. This is one of the most important multiple-component condensation reactions – aside from the Ugi reaction. The separation of products from the starting materials turns out to be relatively difficult here for a normal case. With the fluorous phase technique, however, it can be separated from all impurities and starting materials that have not been converted via simple liquid extraction with suitable fluorous solvents, because the product, in contrast to the starting materials, is bound to a fluorous label.

Several review articles are recommended as further reading matter on liquid phase synthesis techniques [Gravert and Janda, 1997; Ferritto and Seneci, 1998; Harwig *et al.*, 1999; Sun, 1999; Wentworth, 1999].

3.5 FURTHER APPLICATIONS OF COMBINATORIAL LIBRARIES

Combinatorial chemistry has been used for years almost exclusively in pharmaceutical drug development and in related research areas; more recently combinatorial libraries have also come to the attention of other scientific areas (for review see Xiang [1999]):

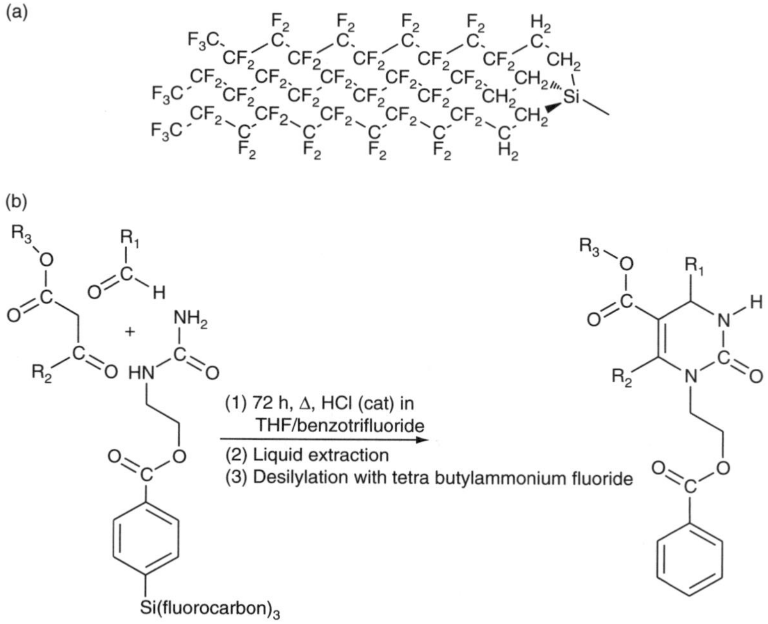

Figure 3.16 Fluorous phases for the synthesis of libraries on a soluble carrier according to Studer *et al.* [1997a,b]. (a) Example of a fluorous label. (b) Application of the strategy with the example of a Biginelli reaction: a urea, an aldehyde and a β-ketoester are converted into diydropyrimidines with this multi-component reaction. The separation of the starting materials, which is normally difficult, can take place through simple solvent extraction of the fluorocarbon-bound products with suitable fluorous solvents.

- superconductors [Xiang *et al.*, 1995];
- magnetoresistive materials [Briceno *et al.*, 1995];
- chemical catalysts [Menger *et al.*, 1995];
- metal chelators [Burger and Still, 1995];
- inorganic luminescence materials [Danielson *et al.*, 1997; Wang *et al.*, 1998];
- sensors [Dickinson *et al.*, 1997];
- dielectrics [Vandover *et al.*, 1998];
- zeolites [Akporiaye *et al.*, 1998];
- electrochemical catalysts [Reddington *et al.*, 1998];
- protein crystallization [Stevens, 2000].

The synthesis of the arrays and libraries takes place here in an analogous fashion to the previously described processes using very diverse materials or material mixtures; the products that arise are tested in turn for the desired properties, such as superconductivity or luminescent behavior, for example.

REFERENCES

Akporiaye, D. E., Dahl, I. M., Karlsson, A. and Wendelbo, R. (1998). Combinatorial approach to the hydrothermal synthesis of zeolites. *Angew. Chem. Int. Ed. Engl.* **37**, 609–611.

Andersson, P. M., Linusson, A., Wold, S., Sjostrom, M., Lundstedt, T. and Norden, B. (1999). Design of small libraries for lead exploration. *Mol. Divers. Drug Des.* 197–220.

Barron, A. E. and Zuckermann, R. N. (1999). Bioinspired polymeric materials: in-between proteins and plastics. *Curr. Opin. Chem. Biol.* **3**, 681–687.

Bayer, E. and Mutter, M. (1972). Liqid phase synthesis of peptides. *Nature* **237**, 512–513.

Briceno, G., Chang, H. Y., Sun, X. D., Schultz, P. G. and Xiang, X. D. (1995). A class of cobalt oxide magnetoresistance materials discovered with combinatorial synthesis. *Science* **270**, 273–275.

Brown, R. D. (1998). Recent developments in solid-phase organic synthesis. *J. Chem. Soc., Perkin Trans. 1* 3293–3320.

Bunin, B. A. and Ellman, J. A. (1992). A general and expedient method for the solid-phase synthesis of 1,4-benzodiazepine derivatives. *J. Am. Chem. Soc.* **114**, 10 997–10 998.

Bunin, B. A., Plunkett, M. J. and Ellman, J. A. (1994). The combinatorial synthesis and chemical and biological evaluation of a 1,4-benzodiazepine library. *Proc. Natl. Acad. Sci. USA* **91**, 4708–4712.

Bunin, B. A., Plunkett, M. J. and Ellman, J. A. (1996). Synthesis and evaluation of 1,4-benzodiazepine libraries. In: Abelson, J. N. (ed.), *Combinational Chemistry* vol. 267, Academic Press, San Diego, pp. 448–465.

Burger, M. T. and Still, W. C. (1995). Synthetic ionophores – encoded combinatorial libraries of cyclen-based receptors for Cu^{2+} and Co^{2+}. *J. Org. Chem.* **60**, 7382–7383.

Carell, T., Wintner, E. A., Bashirhashemi, A. and Rebek, J. (1994a). A novel procedure for the synthesis of libraries containing small organic molecules. *Angew. Chem. Int. Ed. Engl.* **33**, 2059–2061.

Carell, T., Wintner, E. A. and Rebek, J. (1994b). A solution-phase screening-procedure for the isolation of active compounds from a library of molecules. *Angew. Chem. Int. Ed. Engl.* **33**, 2061–2064.

Cheng, S., Comer, D. D., Williams, J. P., Myers, P. L. and Boger, D. L. (1996). Novel solution phase strategy for the synthesis of chemical libraries containing small organic molecules. *J. Am. Chem. Soc.* **118**, 2567–2573.

Cho, C. Y., Moran, E. J., Cherry, S. R., Stephans, J. C., Fodor, S. P., Adams, C. L., Sundaram, A., Jacobs, J. W. and Schultz, P. G. (1993). An unnatural biopolymer. *Science* **261**, 1303–1305.

Coe, D. M. and Storer, R. (1999). Solution-phase combinatorial chemistry. *Annu. Rep. Comb. Chem. Mol. Diversity* **2**, 1–8.

Curran, D. P. (1999). Parallel synthesis with fluorous reagents and reactants. *Med. Res. Rev.* **19**, 432–438.

Danielson, E., Golden, J. H., McFarland, E. W., Reaves, C. M., Weinberg, W. H. and Wu, X. D. (1997). A combinatorial approach to the discovery and optimization of luminescent materials. *Nature* **389**, 944–948.

DeWitt, S. H., Kiely, J. S., Stankovic, C. J., Schroeder, M. C., Cody, D. M. and Pavia, M. R. (1993). 'Diversomers': an approach to nonpeptide, nonoligomeric chemical diversity. *Proc. Natl. Acad. Sci. USA* **90**, 6909–6913.

Dickinson, T. A., Walt, D. R., White, J. and Kauer, J. S. (1997). Generating sensor diversity through combinatorial polymer synthesis. *Anal. Chem.* **69**, 3413–3418.

Felder, E. R. (1999). Solid-phase synthesis of organic libraries. *Comb. Chem. Technol.* 23–33.

Ferritto, R. and Seneci, P. (1998). High throughput purification methods in combinatorial solution phase synthesis. *Drugs Future* **23**, 643–654.

Früchtel, J. S. and Jung, G. (1996). Organic chemistry on solid supports. *Angew. Chem. Int. Ed. Engl.* **35**, 17–42.

Gayo, L. M. (1998). Solution-phase library generation: methods and applications in drug discovery. *Biotechnol. Bioeng.* **61**, 95–106.

Goff, D. A. and Zuckermann, R. N. (1995). Solid-phase synthesis of highly substituted peptoid 1 (2H)-isoquinolinones. *J. Org. Chem.* **60**, 5748–5749.

Gordeev, M. F., Patel, D. V. and Gordon, E. M. (1996). Approaches to combinatorial synthesis of heterocycles – a solid-phase synthesis of 1,4-dihydropyridines. *J. Org. Chem.* **61**, 924–928.

Gordon, D. W. and Steele, J. (1995). Reductive alkylation on a solid phase: synthesis of a piperazinedione combinatorial library, *Bioorg. Med. Chem. Lett.* **5**, 47–50.

Gravert, D. J. and Janda, K. D. (1997). Synthesis on soluble polymers: new reactions and the construction of small molecules. *Curr. Opin. Chem. Biol.* **1**, 107–113.

Hall, S. E. (1999). Recent advances in solid phase synthesis. *Annu. Rep. Comb. Chem. Mol. Diversity* **2**, 15–26.

Han, H., Wolfe, M. M., Brenner, S. and Janda, K. D. (1995). Liquid-phase combinatorial synthesis. *Proc. Natl. Acad. Sci. USA* **92**, 6419–6423.

Harwig, C. W., Gravert, D. J. and Janda, K. D. (1999). Soluble polymers: new options in both traditional and combinatorial synthesis. *Chemtracts* **12**, 1–26.

Kahne, D. (1997). Combinatorial approaches to carbohydrates. *Curr. Opin. Chem. Biol.* **1**, 130–135.

Kaldor, S. W. and Siegel, M. G. (1998). Synthetic organic chemistry on solid support, 307–335. In: Gordon, E. M. and Kerwin, J. F. J. (eds), *Combinatorial Chemistry and Molecular Diversity in Drug Discovery*, John Wiley & Sons, New York.

Keating, T. A. and Armstrong, R. W. (1995). Molecular diversity via a convertible isocyanide in the Ugi four-component condensation. *J. Am. Chem. Soc.* **117**, 7842–7843.

Keating, T. A. and Armstrong, R. W. (1996). Postcondensation modifications of Ugi four-component condensation products: 1-isocyanocyclohexene as a convertible isocyanide. Mechanism of conversion, synthesis of diverse structures, and demonstration of resin capture. *J. Am. Chem. Soc.* **118**, 2574–2583.

Kim, R. M., Manna, M., Hutchins, S. M., Griffin, P. R., Yates, N. A., Bernick, A. M. and Chapman, K. T. (1996). Dendrimer-supported combinatorial chemistry. *Proc. Natl. Acad. Sci. USA* **93**, 10 012–10 017.

Liang, R., Yan, L., Loebach, J., Ge, M., Uozumi, Y., Sekanina, K., Horan, N., Gildersleeve, J., Thompson, C., Smith, A., Biswas, K., Still, W. C. and Kahne, D. (1996). Parallel synthesis and screening of a solid phase carbohydrate library. *Science* **274**, 1520–1522.

Lowe, J. B., Stoolman, L. M., Nair, R. P., Larsen, R. D., Berhend, T. L. and Marks, R. M. (1990). ELAM-1-dependent cell adhesion to vascular endothelium determined by a transfected human fucosyltransferase cDNA. *Cell* **63**, 475–484.

Menger, F. M., Eliseev, A. V. and Migulin, V. A. (1995). Phosphatase catalysis developed via combinatorial organic-chemistry. *J. Org. Chem.* **60**, 6666–6667.

Meyers, H. V., Dilley, G. J., Durgin, T. L., Powers, T. S., Winssinger, N. A., Zhu, H. and Pavia, M. R. (1995). Multiple simultaneous synthesis of phenolic libraries. *Mol. Divers.* **1**, 13–20.

Murphy, M. M., Schullek, J. R., Gordon, E. M. and Gallop, M. A. (1995). Combinatorial organic synthesis of highly functionalized pyrrolidines: identification of a potent angiotensin converting enzyme inhibitor from a mercaptoacyl proline library. *J. Am. Chem. Soc.* **117**, 7029–7030.

Nuss, J. M. and Renhowe, P. A. (1999). Advances in solid-supported organic synthesis methods, 1998 to 1999. *Curr. Opin. Drug Discovery Dev.* **2**, 631–650.

Ostresh, J. M., Husar, G. M., Blondelle, S. E., Dorner, B., Weber, P. A. and Houghten, R. A. (1994). 'Libraries from libraries': chemical transformation of combinatorial libraries to extend the range and repertoire of chemical diversity. *Proc. Natl. Acad. Sci. USA* **91**, 11 138–11 142.

Pei, Y. and Moos, W. H. (1994). Post-modification of peptoid side chains: (3 + 2) cycloaddition of nitrile oxides with alkenes and alkynes on the solid-phase. *Tetrahedron Lett.* **35**, 5825–5828.

Phillips, M. L., Nudelman, E., Gaeta, F. C., Perez, M., Singhal, A. K., Hakomori, S. and Paulson, J. C. (1990). ELAM-1 mediates cell adhesion by recognition of a carbohydrate ligand, sialyl-Lex. *Science* **250**, 1130–1132.

Pirrung, M. C. and Chen, J. (1995). Preparation and screening against acetylcholinesterase of a nonpeptide indexed combinatorial library. *J. Am. Chem. Soc.* **117**, 1240–1245.

Reddington, E., Sapienza, A., Gurau, B., Viswanathan, R., Sarangapani, S., Smotkin, E. S. and Mallouk, T. E. (1998). Combinatorial electrochemistry – a highly parallel, optical screening method for discovery of better electrocatalysts. *Science* **280**, 1735–1737.

Sarshar, S., Siev, D. and Mjalli, A. M. M. (1996). Imidazole libraries on solid support. *Tetrahedron Lett.* **37**, 835–838.

Schweizer, F. and Hindsgaul, O. (1999). Combinatorial synthesis of carbohydrates. *Curr. Opin. Chem. Biol.* **3**, 291–298.

Seeberger, P. H. and Haase, W.-C. (2000). Solid-phase oligosaccharide synthesis and combinatorial carbohydrate libraries. *Chem. Rev.* **100**, 4349–4393.

Simon, R. J., Kania, R. S., Zuckermann, R. N., Huebner, V. D., Jewell, D. A., Banville, S., Ng, S., Wang, L., Rosenberg, S., Marlowe, C. K., Spellmeyer, D. C., Tan, R. Y., Frankel, A. D., Santi, D. V., Cohen, F. E. and Bartlett, P. A. (1992). Peptoids: a modular approach to drug discovery. *Proc. Natl. Acad. Sci. USA* **89**, 9367–9371.

Smith, P. W., Lai, J. Y. Q., Whittington, A. R., Cox, B., Houston, J. G., Stylli, C. H., Banks, M. N. and Tiller, P. R. (1994). Synthesis and biological evaluation of a library containing potentially 1600 amides/esters. A strategy for rapid compound generation and screening. *Bioorg. Med. Chem. Lett.* **4**, 2821–2824.

Sofia, M. J. and Silva, D. J. (1999). Recent developments in solid- and solution-phase methods for generating carbohydrate libraries. *Curr. Opin. Drug Discovery Dev.* **2**, 365–376.

Soth, M. J. and Nowick, J. S. (1997). Unnatural oligomers and unnatural oligomer libraries. *Curr. Opin. Chem. Biol.* **1**, 120–129.

St Hilaire, P. M. and Meldal, M. (2000). Glycopeptide and oligosaccharide libraries. *Angew. Chem. Int. Ed. Engl.* **39**, 1163–1179.

Stevens, R. C. (2000). High-throughput protein crystallization. *Curr. Opin. Struct. Biol.* **10**, 558–563.

Studer, A., Hadida, S., Ferritto, R., Kim, S. Y., Jeger, P., Wipf, P. and Curran, D. P. (1997a). Fluorous synthesis: a fluorous-phase strategy for improving separation efficiency in organic synthesis. *Science* **275**, 823–826.

Studer, A., Jeger, P., Wipf, P. and Curran, D. P. (1997b). Fluorous synthesis – fluorous protocols for the Ugi and Biginelli multicomponent condensations. *J. Org. Chem.* **62**, 2917–2924.

Sun, C.-M. (1999). Recent advances in liquid-phase combinatorial chemistry. *Comb. Chem. High Throughput Screen.* **2**, 299–318.

Suto, M. J. (1999). Developments in solution-phase combinatorial chemistry. *Curr. Opin. Drug Discovery Dev.* **2**, 377–384.

Taylor, C. M. (1997). Strategies for combinatorial libraries of oligosaccharides. In: *Combinatorial Chemistry. Synthesis and Application* (Edited by S. R. Wilson and A. W. Czarnik), Wiley, New York, pp. 207–224.

Thompson, L. A. and Ellman, J. A. (1996). Synthesis and applications of small-molecule libraries. *Chem. Rev.* **96**, 555–600.

Vandover, R. B., Schneemeyer, L. D. and Fleming, R. M. (1998). Discovery of a useful thin-film dielectric using a composition-spread approach. *Nature* **392**, 162–164.

Walz, G., Aruffo, A., Kolanus, W., Bevilacqua, M. and Seed, B. (1990). Recognition by ELAM-1 of the sialyl-Lex determinant on myeloid and tumor cells. *Science* **250**, 1132–1135.

Wang, J. S., Yoo, Y., Gao, C., Takeuchi, I., Sun, X. D., Chang, H. Y., Xiang, X. D. and Schultz, P. G. (1998). Identification of a blue photoluminescent composite material from a combinatorial library. *Science* **279**, 1712–1714.

Wentworth, P. (1999). Recent developments and applications of liquid-phase strategies in organic synthesis. *Trends Biotechnol.* **17**, 448–452.

Xiang, X. D. (1999). Combinatorial materials synthesis and high-throughput screening. An integrated materials chip approach to mapping phase diagrams and discovery and optimization of functional materials. *Biotechnol. Bioeng.* **61**, 227–241.

Xiang, X. D., Sun, X. D., Briceno, G., Lou, Y. L., Wang, K. A., Chang, H. Y., Wallacefreedman, W. G., Chen, S. W. and Schultz, P. G. (1995). A combinatorial approach to materials discovery. *Science* **268**, 1738–1740.

Zuckermann, R. N., Kerr, J. M., Kent, S. B. H. and Moos, W. H. (1992). Efficient method for the preparation of peptoids [oligo(*N*-substituted glycines)] by submonomer solid-phase synthesis. *J. Am. Chem. Soc.* **11**, 10 646–10 647.

4 Chemical Libraries Based on the use of Mixtures

4.1 SYNTHESIS POSSIBILITIES

Using mixture techniques is indispensable for the rapid synthesis of libraries that are to include a number of individual compounds that is as high as possible (see Figure 2.25). In principle, a distinction can be made here between two synthesis variants:

- the coupling of compound mixtures;
- the parallel coupling of individual compounds with subsequent pooling to get polymer-bound product mixtures.

The first variant is significantly simpler and quicker in synthesis, but the equal distribution of the individual compounds can only be realized with more difficulty. This is required, however, in order to be able to compare the activities of the individual compounds with each other within the scope of the biological screening.

The strategy for identification of the product with the 'most interesting' properties also heavily depends on the mixture technique that is selected, thus the individual concepts are to be delved into in more detail below.

4.1.1 THE COUPLING OF COMPOUND MIXTURES

As already mentioned in Section 2.3 as an introduction, this method was implemented by Geysen *et al.* in 1986. They synthesized and analyzed pin-bound peptide mixtures that represented millions of different peptides in order to characterize a discontinuous viral epitope of a monoclonal antibody [Geysen *et al.*, 1986]. They made a distinction with regard to this between 'mixed' positions (referred to as 'X') and 'defined' positions (referred to as 'O') (Figure 4.1). To create a 'mixed' position, a mixture of building blocks (e.g. all 20 naturally occurring amino acids) is coupled instead of a specific building block (e.g. one amino acid). 'Defined' positions are, in contrast, created through a separate coupling of individual building blocks in parallel sets; so-called sub-libraries arise in the process. In the case of Geysen *et al.*, two 'defined' positions of natural amino acids accordingly required $20 \times 20 = 400$ parallel syntheses.

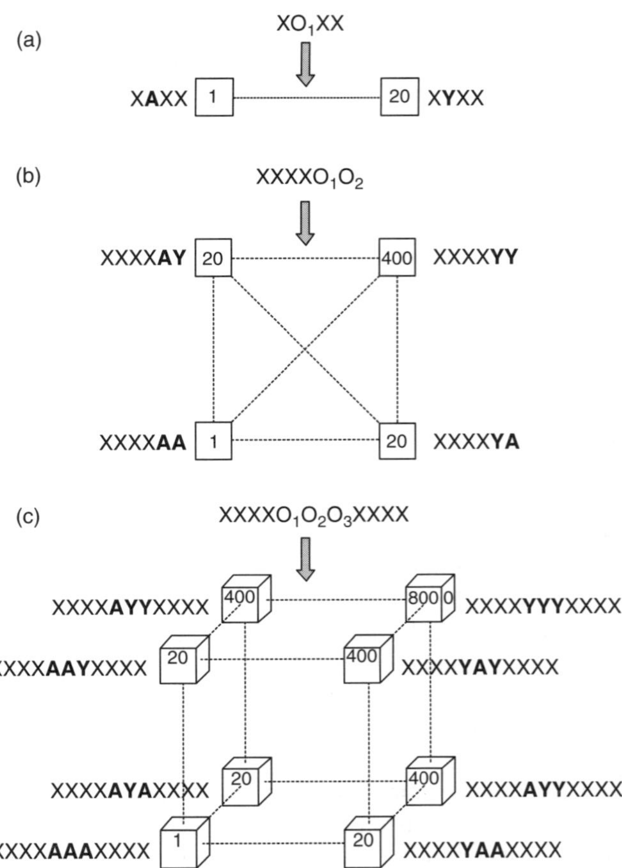

Figure 4.1 Coupling of compound mixtures to create chemical libraries with the example of peptide libraries: 'Mixed' positions (referred to as 'X') are created by coupling mixtures of all 20 natural amino acids; 'defined' positions (referred to as 'O') are created by coupling the natural amino acids in 20 parallel sets (without recombination). (a) A 'defined' position results in 20 so-called sublibraries, each containing $20^3 = 8000$ individual peptides, because three 'mixed' positions are coupled. (b) Two 'defined' positions and four 'mixed' positions result in $20^2 = 400$ sublibraries, each containing $20^4 = 160\,000$ individual peptides. (c) Three 'defined' positions and eight 'mixed' positions result in $20^3 = 8000$ sublibraries, each containing $20^8 = 25.6$ billion individual peptides.

Since they also coupled six 'mixed' positions, each of the 400 sublibraries contained a mixture of $20^6 = 64\,000\,000$ peptides. The number of 'defined' positions consequently determines the number of sublibraries; the number of 'mixed' positions, in contrast, determines the number of individual compounds in each sublibrary. The parallel syntheses were carried out on polyethylene pins; other methods could also be used in principle, for example the tea bag strategy or the spot technique (see Sections 2.2 and 2.3).

The identification of a biologically active compound mixture (or an active individual compound contained within it) then takes place neither through an analysis nor a separation of the mixtures, but instead through a biological test. Geysen *et al.* carried out an enzyme-linked immunosorbent assay (ELISA) with the pin-bound peptide mixtures for this; analogously to this, the process can also be done with membrane-bound compound spots (Figure 4.2).

In this way, the essential building blocks at the 'defined' positions can be determined and become 'fixed' positions (referred to as 'A', of 'C', or 'D' etc, i.e. the single letter code of the essential amino acid). In the subsequent steps 'mixed' positions can be transformed into 'defined' positions and the latter into 'fixed' positions bit by bit while retaining the 'fixed' positions, in order to decode all positions. This iterative process represents the original way of deconvolution, i.e. the derivation of the sequence of the individual compound that shows biological activity (see Section 4.2.3).

To obtain an even distribution of the individual peptides in the mixtures, Geysen *et al.* used an excess of amino acids and varied the concentration of each individual amino acid in accordance with its coupling efficiency. It must be noted here that the coupling efficiency of the 20 naturally occuring amino acids also depends on the amino acid to which they are being coupled. Because there

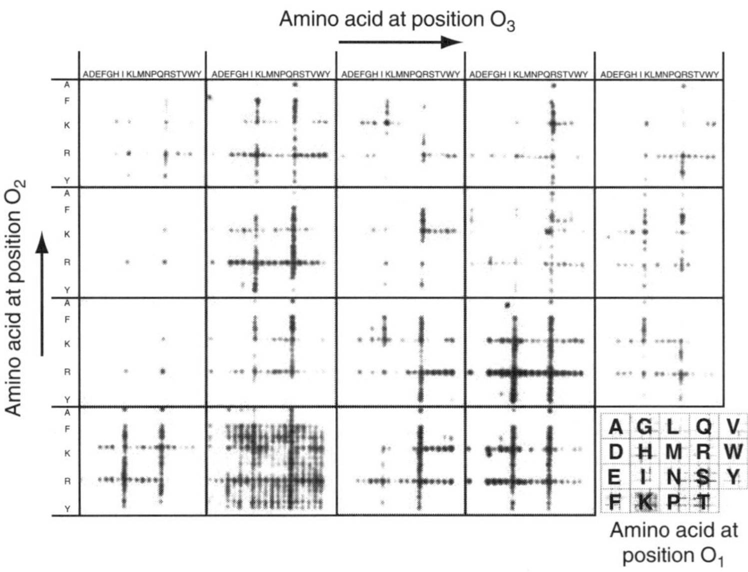

Figure 4.2 Combinatorial peptide library $XXXXO_1O_2O_3XXXX$ after incubation with a target molecule. The library was created with the spot technique on a membrane; the spots represent all possible combinations of L-amino acids at the three 'defined' positions of the peptide. Only those spots that interact with the target molecule show a color reaction (the illustration was kindly made available by Jerini, Berlin, Germany).

are again 20 possibilities for this, $20 \times 20 = 400$ different reaction rates have to accordingly be taken into consideration. They in turn cannot be applied to all of the activation methods without further ado, therefore a relatively great amount of kinetic and mechanistic pre-clarification is necessary for the optimal use of this method [Ivanetich and Santi, 1996].

Another possibility consists in using equimolar amounts of amino acids in the overall ratio of 1:1 to the free coupling positions; although there must be given a complete coupling of all of the amino acids within an appropriate period of time. As an alternative to this, a shortage of the amino acids can also be used – in this case, multiple couplings must accordingly be done.

4.1.2 THE PARALLEL COUPLING OF INDIVIDUAL COMPOUNDS WITH SUBSEQUENT POOLING TO GET POLYMER-BOUND PRODUCT MIXTURES

Portioning–Mixing

Furka *et al.* introduced portioning of polymeric supports for the separate coupling of amino acids at each 'mixed' position with a subsequent pooling of the polymer in 1988, with the main goal of synthesizing more individual peptides within a shorter period of time [Furka *et al.*, 1988a,b; 1991]. The biological screening of peptides was not at the center of attention here.

In principle, the resin is evenly portioned before the coupling of 'mixed' positions (in accordance with the number of amino acids to be coupled). One of the amino acids is then separately coupled in each vessel. This does, in fact, take more time than the use of amino acid mixtures, but equal concentrations of all of the individual peptides can be ensured in the product mixtures because of the competition that is lacking between amino acids that couple well and those that couple poorly. After this, the polymer is once again mixed, washed and the *N*-terminal protecting group is removed. After renewed portioning, the procedure is repeated until the desired length of the peptide has been achieved. The peptides are separated from each other after cleavage from the polymer by means of preparative high-performance liquid chromatography (HPLC), which simultaneously represents the purification. Problems arise if the products have very similar physicochemical properties or if there are a lot of byproducts both of which can no longer be separated using HPLC. This can frequently not be foreseen, especially if several positions are to be exchanged simultaneously and repeatedly. Furthermore, the retention time may also be influenced by secondary structure in the case of larger peptides. The final identification then takes place via sequencing.

Divide, Couple, and Recombine

Houghten *et al.* developed, independently of Furka *et al.*, an approach that is analogous in its fundamentals but that goes much further. The objective was to synthesize mixtures of millions of free peptides in a short period of time in amounts that were so large that their direct screening was able to be carried out in all of the desired assay systems.

Starting from the tea bag synthesis developed by them (see Figure 2.23), they created peptide libraries that provided large amounts of products for biological screening in a relatively short period of time by dividing the polymer into individual tea bags before the coupling of 'mixed' positions, separate coupling of the individual amino acids and subsequent recombination of the polymer (divide, couple, and recombine, DCR) [Houghten *et al.*, 1991]. However, the screening only took place after cleavage of the peptide mixtures from the polymer. This has the advantage that the peptide mixtures can also be used for investigations of membrane-bound proteins, like receptors, or even whole cells.

The determination of the 'most interesting' peptide sequence in the end again took place in an analogous fashion to Geysen *et al.* through iterative deconvolution by means of 'fixed' positions, 'defined' positions, and 'mixed' positions or else with an alternative deconvolution strategy, the so-called positional scanning (see Section 4.2.3).

A comprehensive overview of Houghten *et al.*'s approach was published recently [Houghten *et al.*, 1999].

Split Synthesis or 'One Bead One Compound'

At the same time as Houghten *et al.*, Lam *et al.* developed a further possibility for the synthesis and testing of a large number (>1 000 000) of different peptides that was likewise based on Furka *et al.*'s method [Lam *et al.*, 1991]. In contrast to Houghten *et al.*, however, quick identification of the screened peptide without deconvolution strategies taking a lot of time was at the center of attention here.

Small polymer beads with a diameter of 100–200 μm served as polymeric supports (Figure 4.3); approximately 50–200 pmol peptide per bead was synthesized with the method that has already been described in the previous sections – called split synthesis in this case (Figure 4.4(a)). Every individual bead then only carries peptides with the same sequence then, as a result, that is naturally different from bead to bead, which earned the approach the name 'one bead one peptide'. In principle, this method can be transferred to all compound classes, so the name was changed to 'one bead one compound'. To synthesize a complete library, however, it has to be noted that the number of beads used should be greater than the number of compounds that are to theoretically be expected (10 beads per individual sequence have proved to be useful).

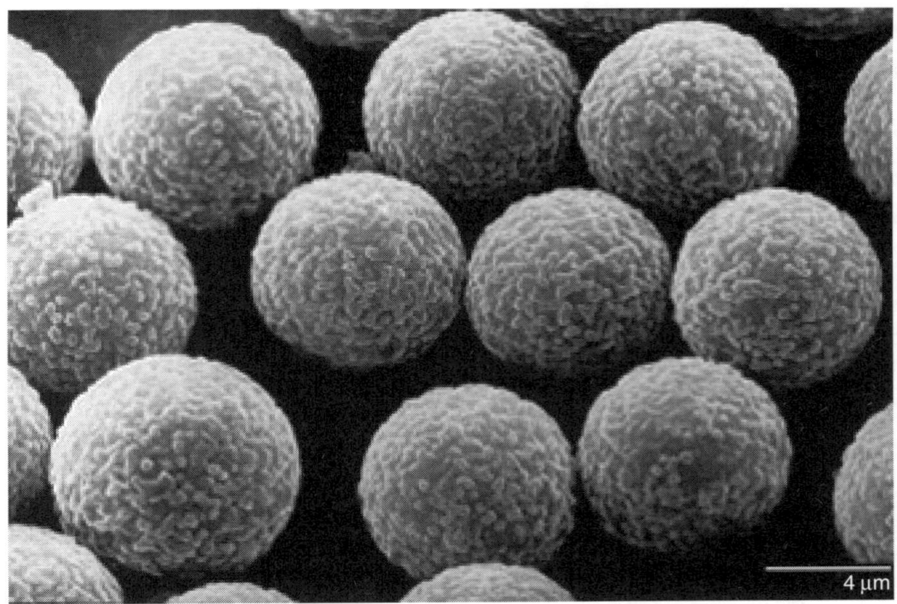

Figure 4.3 TentaGel beads for the synthesis of 'one bead one compound' libraries (the illustration was kindly made available by Rapp Polymere, Tübingen, Germany).

In principle, the identity of the peptides on each bead can be clarified quickly and directly through microsequencing. Since this issue is only of interest after the bead that carries the 'most interesting' peptide sequence has been found with a suitable screening strategy, the peptide has to remain on the bead in the meanwhile, however.

The screening of polymer-bound peptides requires as a prerequisite that the hypothetically free *C*-terminus is not involved in the biological activity of the peptide. Furthermore, the screening method should be visualizable and the whole assay system should work in solution or should be solubilizable (see Figure 4.4(b)). This is generally true for the study of antibody–peptide interactions; this is why this system was used by Lam *et al.* In the meanwhile, various other screening variants have been used, for example radioactive labeling of the binding partners for microautoradiography [Nestler *et al.*, 1996] and direct dye labeling of the binding partners [Chen *et al.*, 1998]. New opportunities resulted from the use of fluorescently labelled secondary antibodies, because the beads can now be automatically sorted with the aid of a fluorescence-activated cell sorter (FACS) [Needels *et al.*, 1993].

Assays in solution can also be successfully made based on the development of suitable linkers with graduated lability. A portion of the peptides is cleaved for screening purpose, the other portion remains covalently bound for subsequent

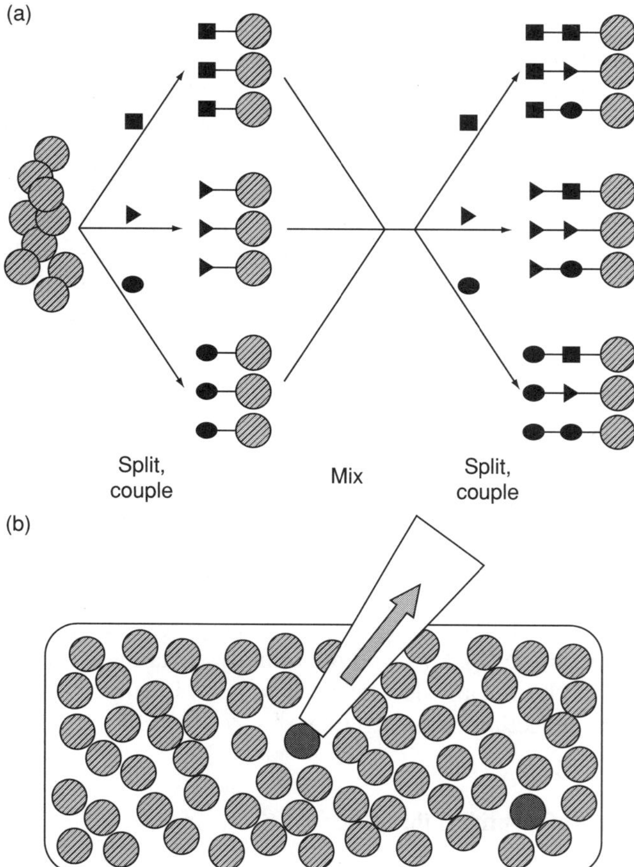

Figure 4.4 'One bead one compound' approach in accordance with Lam *et al.* [1991]. (a) Generation of the library by split synthesis. (b) Screening of the beads. If the peptide sequence of one bead is recognized by an antibody, the peptide–antibody complex, and thereby the bead, can be colored with a second antibody that carries a suitable color-inducing enzyme. The colored peptide-beads are separated from the inactive peptide-beads by a micropipette under the microscope. After this, the sequence of the peptide can be identified through microsequencing.

identification [Salmon *et al.*, 1993]. The beads have to be spatially separated from one another, however, for the screening in solution. One possibility consists in isolating the beads in the wells of a microtiter plate [Salmon *et al.*, 1993]. With the other possibility, the beads have to be immobilized at a certain distance from each other so that the cleaved peptide can only develop its effect in the direct proximity of the bead (diffusion control) [Jayawickreme *et al.*, 1994; Salmon *et al.*, 1996].

Lam *et al.* [1997] and Zhao and Lam [1997] offer comprehensive overviews of the current status of the 'one bead one compound' concept.

4.2 IDENTIFICATION OF THE 'MOST INTERESTING' PRODUCT

Identification of the 'most interesting' compounds of a compound library is often far more demanding than the synthesis and the screening of a library, which is why a number of different strategies have been developed. The most significant are microanalyses, deconvolutions, and encoding strategies, which are to be examined in more detail next. The synthesis of arrays would have to – strictly speaking – be likewise mentioned at this point. But because the identification of the 'most interesting' compounds has already been more or less anticipated in the design of the synthesis of the library and has to only be looked up at the end, this 'comprehensive' concept will be explained separately in Chapter 5.

4.2.1 MICROANALYSIS

Microanalysis is always the first choice whenever possible for the identification of the 'most interesting' compound of a synthetic library. At the moment, three approaches are routinely available: The microsequencing of peptide libraries through Edman degradation, the microsequencing of nucleotide libraries through the dideoxy method according to Sanger (possibly in association with polymer chain reaction (PCR)), and microanalysis via mass spectrometry. The microanalytic methods can also be used here following a so-called affinity selection (see later section of this chapter).

Microsequencing of Peptide Libraries

The high sensitivity of the Edman degradation permits a few pmol of peptide to be sequenced, which is why it is frequently used for the identification of the 'most interesting' sequence in connection with the 'one bead one compound' strategy (Figure 4.5). Furthermore, the so-called pool sequencing has proven to be successful [Stefanovic *et al.*, 1993]. A complete peptide mixture is subjected to Edman degradation here, which permits the quality of smaller libraries to be roughly estimated. In the sequencing of a tripeptide library with the sequence AXV with X = D, G, I, for example, valine would be expected in the first cycle of the Edman degradation, aspartic acid, glycine, and isoleucine in the ratio 1:1:1 in the second cycle, and alanine in the third cycle.

Microsequencing of Oligonucleotide Libraries

Considerably fewer oligonucleotide quantities are accessible for sequencing than is the case with the Edman sequencing of peptides because of the high

Figure 4.5 Edman degradation: In the first step, phenylisothiocyanate (PITC) derivatizes the free *N*-terminus of a peptide under basic conditions; a phenylthiocarbamyl peptide (PTC peptide) is formed. After this, the sulfur performs a nucleophilic attack on the carbonyl carbon of the *N*-terminal amino acid in acidic, anhydrous conditions. As a result, the *N*-terminal peptide is shortened by an amino acid and the relatively unstable anilinothiazolinone amino acid (ATZ amino acid) is formed. Following this, the ATZ amino acid is hydrolyzed into the phenylthiocarbamyl amino acid (PTC amino acid), which subsequently rearranges into the stabile phenylthiohydantoin amino acid (PTH amino acid) through an acid-catalyzed reaction. The PTH amino acid can then be identified and quantified chromatographically.

sensitivity of the deoxyribonucleic acid (DNA) sequencing, which is increased even more by the combination with PCR (Figure 4.6). The DNA sequencing is of major significance in the area of biotechnological processes for creating peptide or protein libraries (see Chapter 6), but also in the direct analysis of oligonucleotide libraries [Ellington and Szostak, 1990; Tuerk and Gold, 1990] and the encoding of libraries by means of nucleotides (see Section 4.2.2).

Mass Spectrometry

The microanalysis of libraries by means of mass spectrometry (MS) is not tied either to the type of individual building blocks or to the type of their linkage and can consequently be universally used for all libraries. It will be explored comprehensively in Section 7.4, because it represents one of the most important analytical methods in the area of combinatorial chemistry.

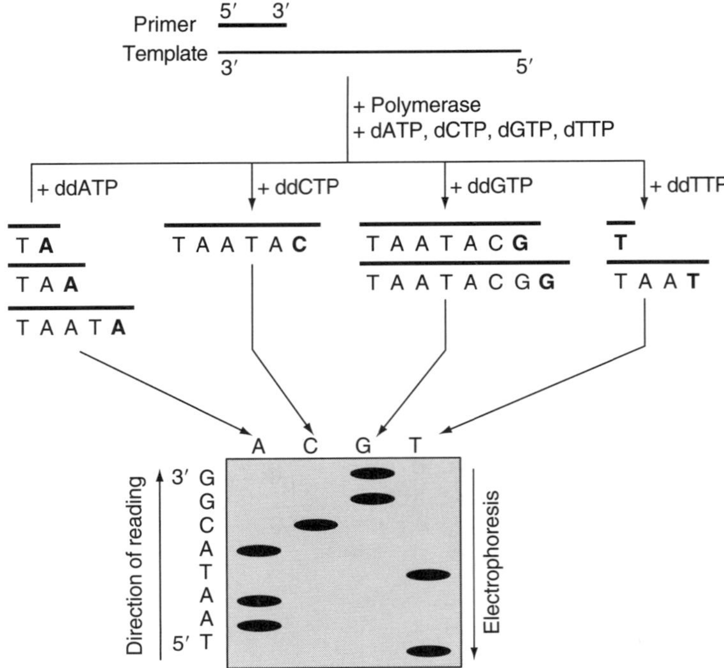

Figure 4.6 Sanger dideoxy sequencing of oligonucleotides: A primed DNA synthesis reaction catalyzed with polymerase is carried out in four parallel sets; in addition to the four 2′-deoxynucleotides dNTP, one radioactive labeled 2′, 3′-dideoxynucleotide ddNTP is added, the incorporation of which leads to a base-specific chain termination. The DNA fragments of different lengths that are formed because of this result in a characteristic band pattern after parallel separation by polyamide gel electrophoresis, from which the sequence can be directly deduced.

Affinity Selection of Libraries

The central part of the affinity selection of molecules from a library is a special screening of the entire mixture; the strength of the interaction with a binding partner is made use of with regard to this to directly separate out the 'most interesting' compounds, meaning compounds with high affinity, from the mixture. The whole library, which can only be accessed with difficulty analytically, is ideally reduced to an individual compound or to a few individual compounds of a sequence motif here, which can then subsequently be subjected to microanalysis.

The method was originally used for oligonucleotide libraries by making use of the affinity of the individual strands for complementary strands for selection [Lew and Kemp, 1989]. Furthermore, it was an important part of the first molecular biological approaches for creating peptide or protein libraries;

antibodies or (fragments of) receptors usually served as binding partners here (see 'phage display', Section 6.1). This method also very quickly came into use for chemically synthesized peptide libraries [Zuckermann *et al.*, 1992; Songyang *et al.*, 1993]. The binding partners (again antibodies or receptors) are first incubated with a peptide excess here. After this, unbound peptides or peptides with low affinity are separated. The high-affinity ligands that are selected with this are then washed away under somewhat drastic conditions and subsequently analyzed; MS methods [Kelly *et al.*, 1996; Kaur *et al.*, 1997] are usually used – Edman pool sequencing is used less frequently [Songyang *et al.*, 1993] (Figure 4.7).

The so-called affinity capillary electrophoresis (ACE) represents an interesting variant [Chu *et al.*, 1993]. The migration behavior of a ligand changes in general after binding to its target molecule, because the electrophoretic mobility of a compound is dependent on its net charge and its mass. These properties can now be made use of for selection by comparing the electropherograms of a library with and without the addition of target molecules (Figure 4.8). Individual, high-affinity ligands from libraries consisting of several hundred peptides were even able to be successfully selected with regard to their binding affinity for the model receptor vancomycin through on-line MS coupling [Chu *et al.*, 1996].

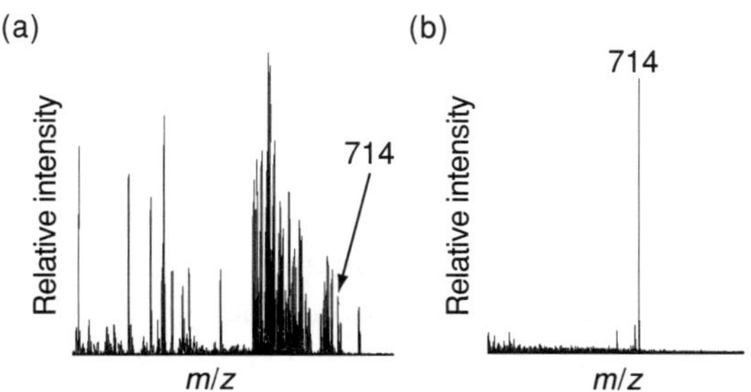

Figure 4.7 Mass spectrum of a library (a) before and (b) after affinity selection (the illustration was kindly made available by Novartis, Basel, Switzerland).

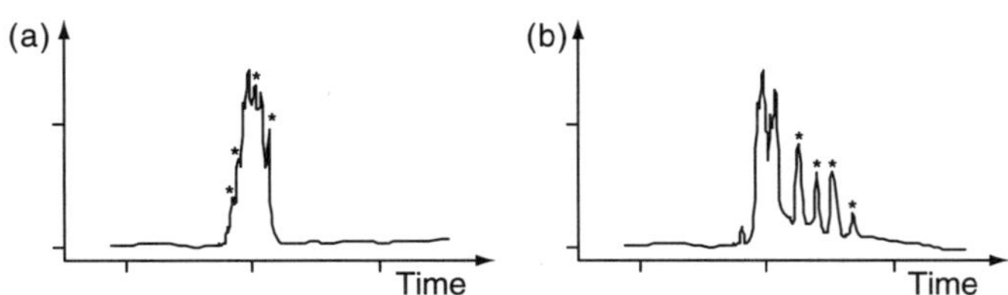

Figure 4.8 Affinity capillary electrophoresis of a peptide library (UV detection) following Chu et al. [1993]. (a) Without receptor and (b) with receptor.

Very promising nuclear magnetic resonance (NMR) experiments have also been developed lately with which affinity selection can be arranged for (see Chapter 7, Section 7.3.3). The high-affinity ligands are not physically separated here, but their signals, which change through the interaction with the target molecule, are merely filtered out from the multitude of other signals.

4.2.2 ENCODING/DECODING

It frequently happens that no system of analysis has been established for the members of a combinatorial library with which conclusions can be made about the identity of the 'most interesting' member in the subsequent structure determination. This applies above all to organic libraries of low molecular weight compounds and to nonnatural polymer libraries that are created using the 'one bead one compound' strategy. Various encoding strategies have been developed to nevertheless make their analysis possible (for review see Eckes [1994], Janda [1994], Jacobs and Ni [1998], Barnes and Balasubramanian [2000] and Tan and Burbaum [2000]). Encoding means the introduction of so-called tags, i.e. identification labels, which translate all of the desired information into one 'language' or code and the analysis of which is possible in a quick and simple manner. A distinction is made between chemical and so-called noninvasive methods. In the former, chemical compounds are used as tags; in the latter in contrast, external storage units such as microchips are used.

Chemical Encoding

Compound classes that allow fast and easy trace analysis – mainly (oligo)nucleotides (PCR amplification with subsequent sequencing), amino acids or peptides (Edman degradation or mass spectrometry), haloaromatic compounds (gas chromatography) and secondary amines (HPLC after derivatization) – are used for the chemical encoding of libraries. A prerequisite for this type of encoding is that there is no interference of the tag and ligand synthesis because coupling of the tags takes place to the same polymeric support as the synthesis of the ligand library. In principle, there exist several types of chemical encoding strategies that are explained in more detail next.

Direct encoding

Direct encoding is above all used in the case of linear oligomer libraries, because every coupling of a synthetic building block is encoded by the sequential coupling of a corresponding tag (Figure 4.9). An *n*-mer ligand sequence is consequently represented by an *n*-mer tag sequence; the structural information of the ligand is encoded by the tags per se and the order of the tags on the linear tag oligomer and can be derived by its sequencing.

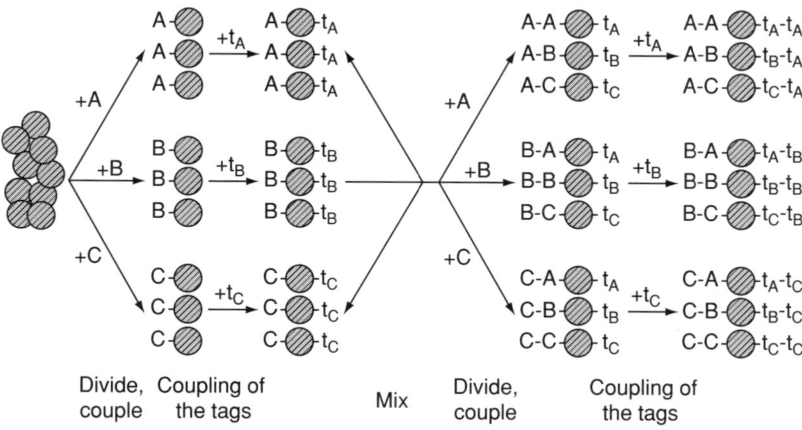

Figure 4.9 Schematic diagram of the direct encoding of libraries by so-called tags. The coupling of a tag ($t_A, t_B, \ldots$), which documents the coupling event in place of the actual building block, takes place after (or even before) every coupling of a library building block (A, B, ...), because it is more capable of being analyzed.

Oligonucleotide tags

The use of DNA to encode libraries was proposed by Brenner and Lerner, because even the smallest quantities of code suffice to exactly sequence them after their PCR amplification [Brenner and Lerner, 1992]. The protecting group strategies within the scope of the chemical DNA synthesis can, however, be influenced in a negative way relatively easily by other chemical reactions; the use of this method remains mainly limited to the encoding of peptide libraries that are based on 9-fluorenylmethoxycarbonyl (Fmoc) chemistry because of this.

The amino acids of an eight-membered tripeptide library of the type $X_3X_2X_1$ (X is limited to Gly and Met) were encoded in each case with hexanucleotide tags within the scope of the first thought experiment, so the analysis of 18mer oligonucleotides would have been necessary for decoding; they would have to be flanked by corresponding primer sequences in the process [Brenner and Lerner, 1992]. The practical implementation took place a little while later with the use of a bifunctional serine anchor [Nielsen *et al.*, 1993] (Figure 4.10). The synthesis of the peptides and encoding nucleotides took place in accordance with Figure 4.9; the results of the alternating synthesis were able to be significantly improved with the introduction of a spacer on the serine anchor (Figure 4.10).

One disadvantage of the method is that the ligand cannot be separated from its code before the screening, hence the result of the subsequent biological assays could be falsified by the coding sequence, although this was not the

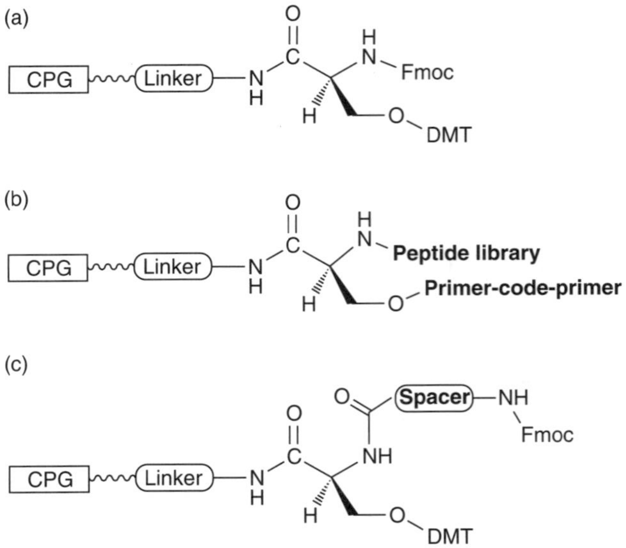

(a)

(b)

(c)

Figure 4.10 Direct encoding of a peptide library by oligonucleotides according to Nielsen *et al.* [1993]. (a) A bifunctional serine anchor that is bound through a linker to porous glass beads (controlled-pore glass, CPG) serves for the parallel synthesis of the two oligomers. (b) The peptide oligomers were synthesized at the amino group of the serine using Fmoc strategy, the encoding oligonucleotides were synthesized with the phosphoramidite method at the side chain of the serine (DMT = dimethoxytrityl protecting group). (c) The synthesis was able to be significantly improved with the introduction of a spacer, because the mutual steric hindrance of the growing oligomer chains was able to be reduced because of this.

case in the example investigated. This problem can be simply mitigated, however, by a different anchor strategy (see below).

In parallel with the first experimental work of Nielsen *et al.*, Needels *et al.* carried out work that was much more comprehensive in the area of nucleotide encoding [Needels *et al.*, 1993]. They succeeded in encoding the $7^7 = 823\,543$ individual compounds of a 'one bead one compound' library of the type $X_7X_6X_5X_4X_3X_2X_1$ (in which X = Arg, Gln, Phe, Lys, Val, D-Val, Thr) by assigning the amino acids to dinucleotide tags in each case and coupling them in an alternating fashion in accordance with Figure 4.9. They did not use bivalent anchors on the polystyrene beads (diameter 10 μm), but instead used different anchors in different ratios for the peptide and nucleotide synthesis (Figure 4.11). Fluorescently labeled antibodies were used for screening of the binding affinity of the peptides for antibodies, so the beads that carried the peptides with the best binding properties were able to be selected by means of FACS. After this, the 14mer nucleotide code – which was naturally flanked again by the corresponding primers – was amplified with PCR and subjected to sequencing, so conclusions were able to clearly be made on the corresponding individual peptides.

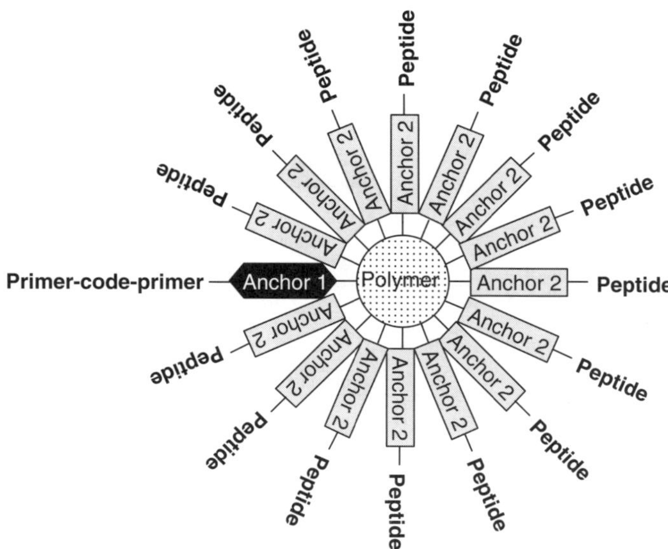

Figure 4.11 Use of different anchors for the direct encoding of libraries according to Nielsen *et al.* [1993]. Instead of bivalent anchors, different anchors are used for the peptide and nucleotide synthesis; the anchors for the peptides are used to a great excess, because substantially more compound is required in general for the biological assays than for the analysis of the DNA code. Furthermore, the probability that the code will influence the result of the biological screening is minimized with this measure.

Amino acid or peptide tags

If peptide libraries are synthesized that contain unnatural amino acids or are *N*-terminally modified, the microanalysis of the library members by Edman degradation, which is otherwise customary, is no longer possible. Amino acid or peptide tags that can be sequenced suggest themselves for the encoding of these libraries, because a series of orthogonal protecting group strategies have been established for peptide synthesis and the compatibility of tag and ligand synthesis can consequently be ensured.

The technique was applied for the first time for the encoding of two 'mixed' positions X_1 and X_2 of an *N*-terminally acetylated decapeptide library of the form $Ac - RAOHTTGX_2IX_1 - NH_2$; O and X essentially included nonnatural amino acids [Kerr *et al.*, 1993]. Tripeptides were used for this as tags, so X = ornithine was represented by the sequence GAF and X = norvaline by the sequence ALG, for example. The sequencing result H-GAFALG-OH accordingly represented a decapeptide with X_2 = ornithine and X_1 = norvaline. The parallel synthesis of the deca- and encoding peptide took place in accordance with Figure 4.9 using a bifunctional Lys anchor. The decapeptide was then synthesized by means of the Fmoc strategy at the N^{α} of the Lys anchor, whereas the tag peptide was synthesized on the N^{ε} of the Lys anchor (Figure 4.12).

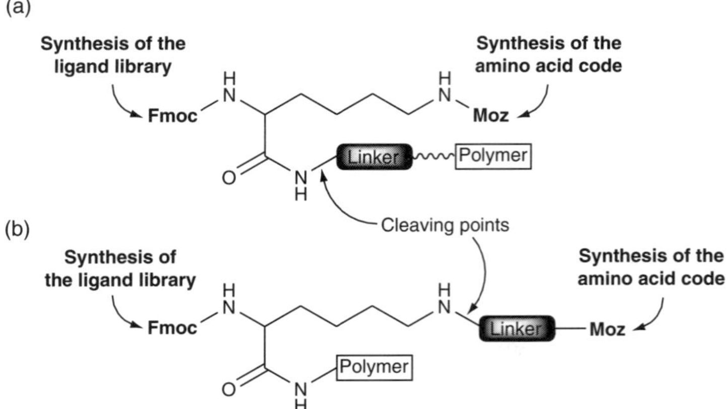

Figure 4.12 Direct encoding of a nonsequenceable peptide library by means of sequenceable peptides according to Kerr *et al.* [1993]. The synthesis is done on a bifunctional lysine anchor by means of Fmoc (N^{α}, piperidine cleavage) and 4-methoxybenzyloxycarbonyl (Moz) α,α-dimethyl-3,5-dimethoxybenzyloxycarbonyl (Ddz) strategy (N^{ε}, cleavage with 5% trifluoroacetic acid (TFA). In the case of (a) the ligand can be separated from the polymer but not from its code, which leads to effects on the screening. This problem can be bypassed by variation of the linker position (b).

As already described above, the ligand can only be separated from the polymer, but not from its code, which may influence the biological assay under certain circumstances. This fundamental problem can either be avoided by variation of the anchor strategy (Figure 4.11) or variation of the linker position (Figure 4.12), as was able to be demonstrated by Nikolaiev *et al.* with the example of the encoding of a library of low molecular weight compounds of the type $X_1X_2X_3$ through individual amino acids as tags [Nikolaiev *et al.*, 1993]. Furthermore, they showed that the encoding of a peptide library using individual amino acid tags is also possible by a combination of the Fmoc strategy (ligand sequence) and the *tert*-butyloxycarbonyl (Boc) strategy (tag sequence). They used an anchor system based on two lysines for this in order to obtain twice as much ligand as code (Figure 4.13).

Binary encoding using molecular tags

For binary encoding of libraries, which is presented in a schematic form in Figure 4.14, information about the ligands is exclusively encoded by the presence and absence of the tags. This is to be illustrated with an example below.

For building up a library with a three-step reaction (in general an *n*-step reaction), 7 compounds shall be used for each of the 3 (n) steps, so that $7^3(7^n)$ different products are to be expected. The different compounds are numbered in a binary form to prepare for the binary encoding: Compound 1 becomes '001', Compound 2 becomes '010', Compound 3 becomes '011',..., and

Synthesis of the ligand library

Fmoc NH Fmoc NH O NH

Synthesis of the amino acid code

Boc H N Linker Polymer O N H

Figure 4.13 Encoding of a ligand library with individual amino acid tags by a combination of the Fmoc strategy and the Boc strategy according to Nikolaiev *et al.* [1993]. The deprotection of the Fmoc groups with 50 % piperidine takes place orthogonally to the deprotection of the Boc groups with 30 % TFA.

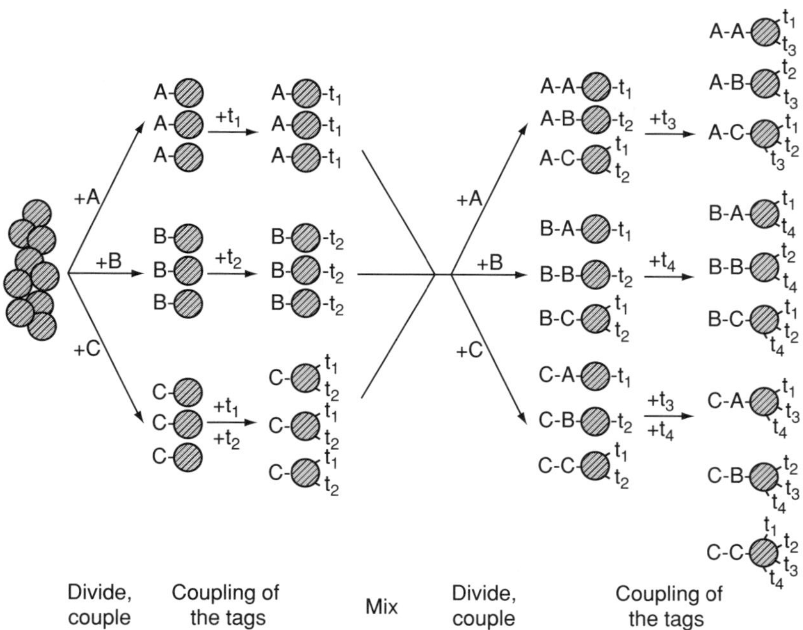

Figure 4.14 Schematic diagram of the binary encoding of libraries. The coupling of binary tags ($t_1, t_2, \ldots$) takes place after (or even before) each coupling of a library building block (A, B, ...). Binary means in this context, that the information about the coupling events is represented by the presence and absence of these tags.

Compound 7 becomes '111'. If one wants to use more than 7 compounds, the binary code has to be expanded by the corresponding number of digits. Generally speaking: $2^m - 1$ compounds can be numbered with an m-digit binary code. To describe an arbitrary n-step synthesis it suffices to set up a binary synthesis code that consists of $m \times n$ binary characters, so $3 \times 3 = 9$ characters in the example chosen. If Compound 2 is coupled in an initial reaction step, for example, then the corresponding binary description of this event is '010'. If Compound 7 is then coupled in the second reaction step, the binary form is '111 010'. The subsequent coupling of Compound 5 would then correspond to '101 111 010'. The representation of this 9-digit binary code in a chemical form requires 9 tags $t_1 - t_9$, which can be differentiated in a simple way and which are easily analyzed. The number 1 is to be translated as 'tag present' and the number 0 with 'tag absent' in the process. The above code '101 111 010' has to be translated with 'don't couple t_1, couple t_2, don't couple t_3, couple t_4, ..., couple t_9', because the count is done from right to left in accordance with the direction of the synthesis. Furthermore, the tags are numbered in accordance with their retention times for reason of clarity (Figure 4.15).

Libraries that are much larger can be encoded with substantially fewer tags through binary encoding than is the case with direct encoding. In addition, the tags do not absolutely have to be bivalent, because no sequential tag oligomers have to be built up. Compound classes can therefore be used that are considerably more chemically stable than amino acids or nucleotides. This is of importance, e.g. for libraries that require relatively drastic synthesis conditions like libraries of low molecular weight organic compounds.

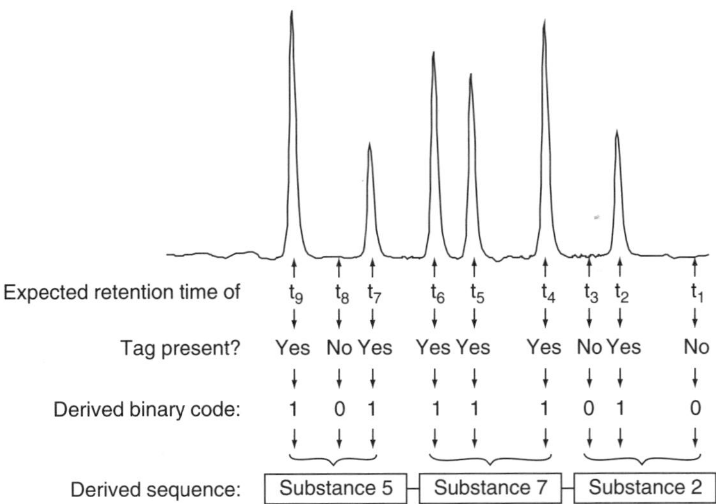

Figure 4.15 Schematic diagram of the analysis of binary tags. In the case that haloaromatic compounds are used, this takes place by means of electron capture gas chromatography. In principle, the code is represented by the presence or absence of the tags.

The concept was implemented for the first time in the work group of Still. Haloaromatic compounds served as tags, because they are very well suited to binary encoding, on the one hand, and are very robust chemically and further can be detected in the subpicomolar range, on the other hand (Figure 4.16) [Ohlmeyer *et al.*, 1993; Borchardt and Still, *1994*; Nestler *et al.*, 1996]. Introduction of the tags can be done specifically through the modification of the polymer or else through the – at first sight – undesired nonspecific modification of both polymer and ligands. Since such a small number of tag molecules are used that only a negligibly small proportion of the ligand molecules would be affected by this, the latter possibility represents a real alternative.

The tags are to be cleaved from the polymer or ligand before the analysis of the code, which takes place in the case of tag type a in Figure 4.16 through irradiation of the UV-sensitive linker with light of an appropriate wavelength and in the case of tag type b, in contrast, through oxidative cleavage by means of ceric ammonium nitrate. The high-sensitivity analysis of the tags through electron capture gas chromatography then takes place after the silylation of the alcohols that arise (see Figure 4.15).

Very stable secondary amines were made use of by Affymax as an alternative to binary encoding by means of haloaromatic tags (Figure 4.17) [Ni *et al.*, 1996]. NH_2–functionalized TentaGel resin was used here, which was equipped with two different linkers in the ratio 9:1. An acid-labile, Fmoc-protected alcoxy-benzyl ester linker for synthesis of the libraries and a Boc-protected glycine

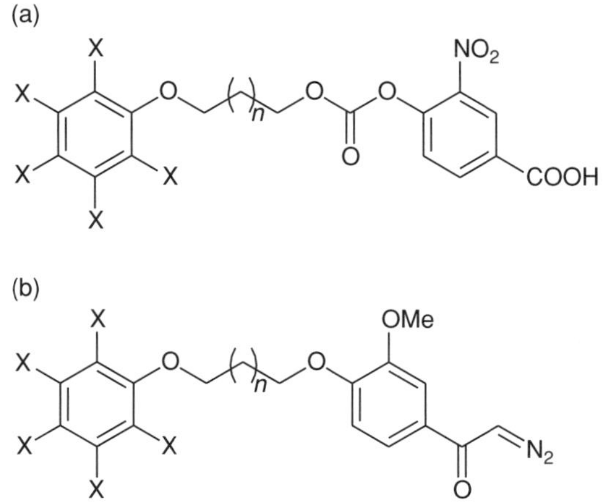

(a)

X X X X X O n O O O NO2 COOH

(b)

X X X X X O n O OMe O N2

Figure 4.16 Example of haloaromatic compounds for the binary encoding of libraries according to Ohlmeyer *et al.* [1993], Borchardt and Still [1994] and Nestler *et al.* [1996]. A large number of different tags can be created by the variation of the halogen atoms X and the length of the hydrocarbon chain *n*. Tag type (a) suggests itself for the encoding of peptide libraries, because the free amino groups of the growing amino acid chains can be modified with this; tag type (b) can, in contrast, be universally used, because it can react with the polystyrene resin or the ligand molecules through rhodium-catalyzed carbene insertion.

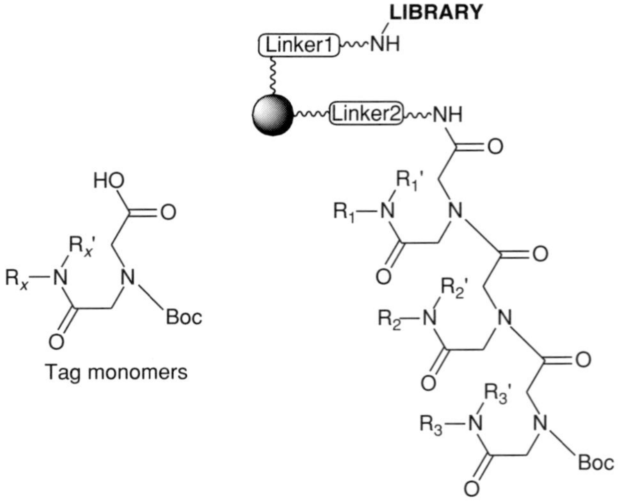

Figure 4.17 Secondary amines for the binary encoding of libraries according to Ni *et al.* [1996]. *N*-Boc-protected *N*-(dialkylcarbamoylmethyl-)glycines serve as tag monomers. The code is created by coupling binary mixtures of the tag monomers in each case by means of *O*-(7-azabenzotriazol-1-yl)-1,1,3,3-tetramethyluronium (HATU) and *N,N*-diisopropylethylamine (DIEA) to the tags of the previous synthesis cycle.

amide linker for synthesis of the codes. After the biological screening and selection of the 'most interesting' ligands, the corresponding tag chains were decomposed into the secondary amines through 15 h of boiling in 6 M HCl at 130 °C. They were subsequently modified with dansyl chloride and, after this, analyzed via HPLC. As schematically depicted in Figure 4.15, the binary code could be directly retrieved from the resulting chromatogram.

Isotopically labeled amino acid or peptide tags

The encoding of libraries with isotopically labeled tags was introduced by Geysen [Geysen *et al.*, 1996]. It is based on chemically identical amino acid or peptide tags that are physically different with regard to their masses, because they have a different distribution of isotopes. After cleavage from the polymer, this mass code can be analyzed very simply and quickly using mass spectrometry (Figure 4.18).

Dyes as tags

The use of dyes as tags – whether through the use of colored glass beads or the modification of polymer beads with colored or fluorescent molecules – represents a very simple type of encoding, because the desired information can be read at any time with the naked eye or under the (fluorescence) microscope [Câmpian *et al.*, 1994; Egner *et al.*, 1997; Guiles *et al.*, 1998]. Only very small

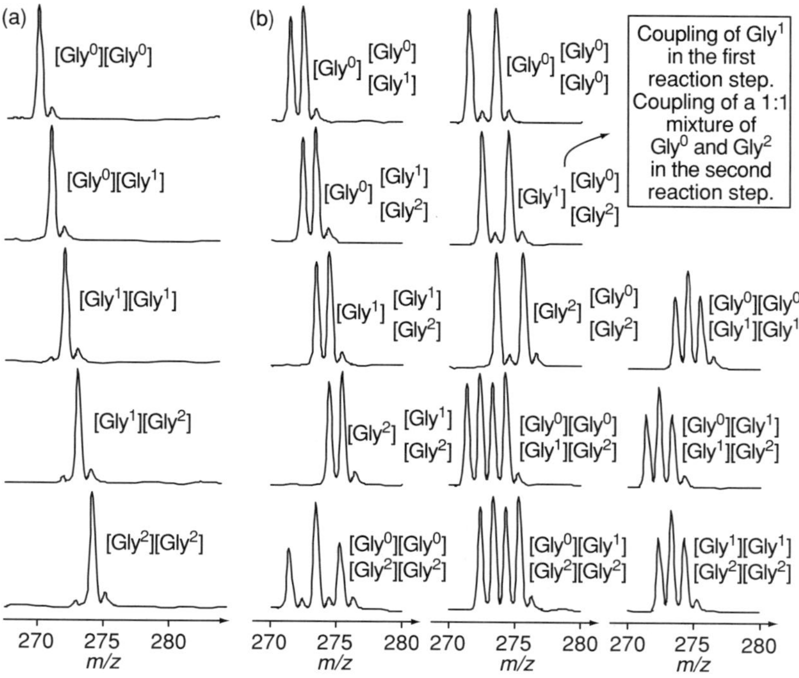

Figure 4.18 Isotopically labeled amino acid or peptide tags for the encoding of libraries following Geysen *et al.* [1996] ([Gly^n] represents a glycine with a mass that is increased by n atomic mass units). (a) Individual peak encoding, i.e. that the tags distinguish themselves in each case by a defined mass. (b) 'Bar code' encoding, i.e. that the tags distinguish themselves in each case by a mass pattern that can be created by coupling isotopically labeled amino acid mixtures (glycine in the case at hand).

libraries have been able to be encoded up to now because internal quenching effects (energy transfer, reabsorption) usually arise after the coupling of various dyes that distort the color information.

Noninvasive Encoding

All approaches of chemical encoding have a more or less strong effect on the chemistry of the libraries to be encoded and require additional syntheses and analytical efforts. Various noninvasive encoding techniques have been developed lately to bypass these problems.

Radio-frequency encoding

The so-called radio-frequency encoding was introduced by two work groups independently of one another in 1995 [Moran *et al.*, 1995; Nicolaou *et al.*, 1995].

The synthesis of the library takes place here in microreactors (Figure 4.19). They contain a microchip as the core element on which radio-frequency signals can be stored that take on the role of the tags. The encoding takes place in accordance with the direct principle shown in Figure 4.9. The radio-frequency signals can be called up again at any time and decoded for structural clarification. The method can, however, only be made use of for the encoding of libraries that are fairly small – but with greater yields – based on the size of the microreactors, because the microreactors are more or less treated as individual polymer beads within the scope of the 'one bead one compound' method (for review see Xiao and Nova [1997]).

Laser optical encoding

Laser optical encoding represents a different variant of the noninvasive encoding of libraries [Xiao *et al.*, 1997]. The synthesis of the library takes place here on so-called laser optical synthesis chips (LOSCs). They are constructed out of a chemically inert, ceramic base and a polypropylene–polystyrene copolymer as the polymeric support (Figure 4.20). Bar codes that are burned into the ceramic bases with the aid of a CO_2 laser serve as tags. The bar code pattern can be read and decoded with the aid of a camera and the corresponding pattern recognition software for structural identification. The method can again only be drawn upon to encode fairly small libraries on the grounds of the chip size, analogously to the radio-frequency encoding.

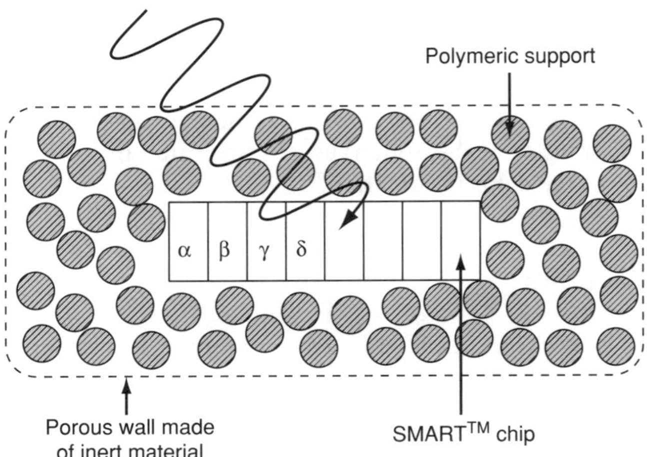

Figure 4.19 Radio-frequency encoding of libraries according to Moran *et al.* [1995] and Nicolaou *et al.* [1995]. The synthesis of the library takes place in microreactors, which are essentially constructed out of a small semiconductor chip (for example a so-called single or multiple addressable radio-frequency tag, SMART™) and the polymeric support. The encoding takes place via radio-frequency signal tags (α, β, γ, δ), which are stored on the microchips.

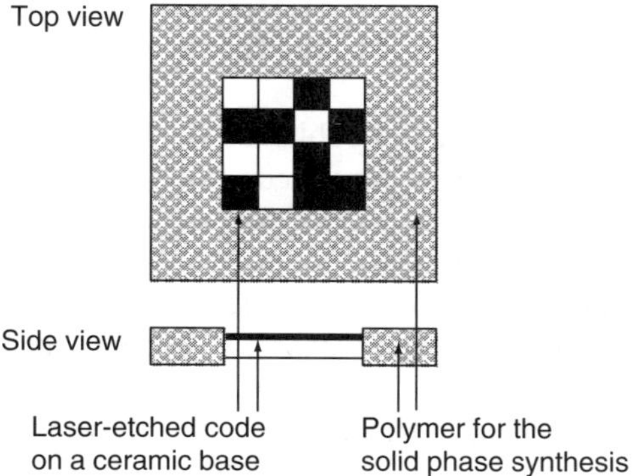

Figure 4.20 Laser encoding of libraries according to Xiao *et al.* [1997] by means of so-called laser optical synthesis chips (LOSCs).

4.2.3 DECONVOLUTION

Iterative Deconvolution

Iterative deconvolution (see also Section 4.1.1) means the step-by-step decoding of the 'most interesting' compound of a library; the following work steps have to be repeatedly run through in the process (Figure 4.21):

- synthesis of a series of compound mixtures with 'defined' positions;
- selection of the 'most interesting' mixture by screening;

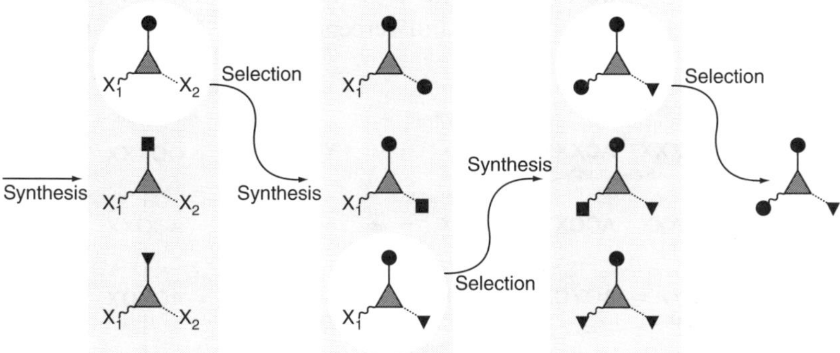

Figure 4.21 Schematic diagram of iterative deconvolution, i.e. of the step-by-step decoding of the 'most interesting' compound. The 'most interesting' mixture is determined by screening of a series of compound mixtures (X = 'mixed' positions) that are distinguished from each other in a 'defined' position. Sublibraries are synthesized taking this as a starting point, the 'most interesting' mixture is again selected and so on, until the 'most interesting' individual compound remains in the end.

- synthesis of sublibraries of the 'most interesting' mixture;
- selection of the 'most interesting' sublibrary by screening.

The number of compounds in the sublibraries becomes smaller and smaller because of this, so that only a single compound remains in the end. The technique was applied by Geysen *et al.* for the first time for peptides [Geysen *et al.*, 1986] and perfected by Houghten *et al.* [1991]. In principle, iterative deconvolution can also be used for other compound classes, for example for oligonucleotide libraries [Ecker *et al.*, 1993] or libraries of low molecular weight organic molecules [Carell *et al.*, 1994; Gordeev *et al.*, 1996]. It is to be explained in more detail next with the example of a pentapeptide (Figure 4.22).

The starting point is the synthesis of sublibraries of the pentapeptide library, which shall comprise three identical 'mixed' positions 'X' (see Section 4.1.1), because they include mixtures of the existing 20 natural amino acids. At the two remaining positions all 20 natural amino acids are also coupled separately in each case to get two 'defined' positions 'O' (see Section 4.1.1), hence $20 \times 20 = 400$ sublibraries have to be synthesized.

If amino acid mixtures are used to create 'mixed' positions, there are no limitations as to the position at which mixed or 'defined' positions have to be located. If the 'mixed' positions are created through portioning and subsequent mixing, it is by far more convenient if the 'defined' positions are located at the *N*-terminus.

After the biological screening of all of the sublibraries, the most active one of them is selected. The two amino acids at the 'defined' positions of this sublibrary are never changed again (they become so-called 'fixed' positions) and serve as the basis for the synthesis of new, smaller sublibraries. One of the three 'mixed' positions is transformed into a 'defined' position here, so 20 new sublibraries have to be synthesized and screened. After this, the most active

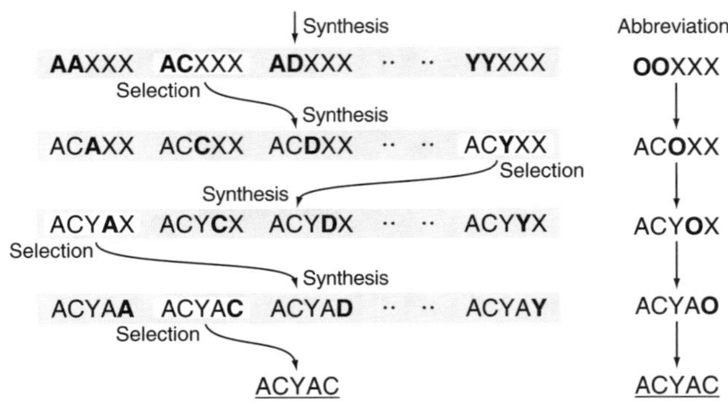

Figure 4.22 Iterative deconvolution with the example of the deconvolution of a pentapeptide.

sublibrary is again selected and the corresponding amino acid at the 'defined' position is transformed into a 'fixed' amino acid. One of the two still remaining 'mixed' positions is now transformed again into a 'defined' position for the renewed synthesis of sublibraries based on this. After renewed screening and subsequent selection, the remaining 'mixed' position is likewise transformed into a 'defined' position, in order to get to the ultimately active sequence in a last cycle.

Positional Scanning

In this deconvolution strategy, the 'most interesting' compound is decoded in a single step by determining the most active individual building block at every position in parallel (Figure 4.23) [Jung and Beck-Sickinger, 1992; Pinilla *et al.*, 1992]. This is to be explained in more detail with the example of a pentapeptide (Figure 4.24).

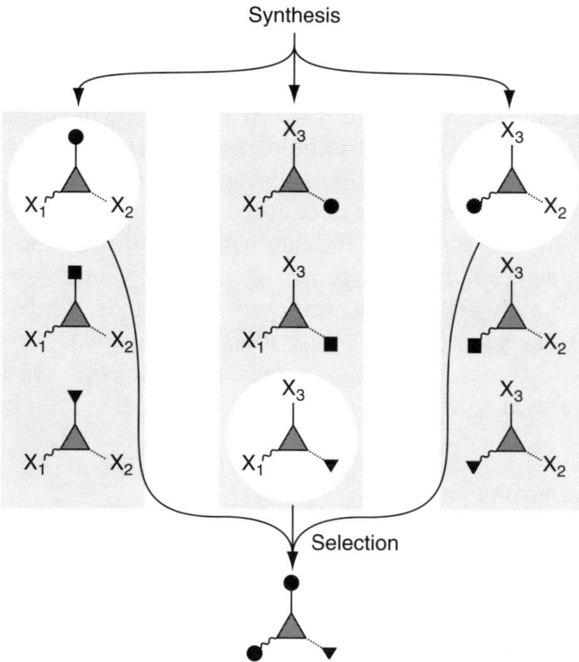

Figure 4.23 Schematic diagram of positional scanning, i.e. of the parallel decoding of the 'most interesting' compound in a single step. Sublibraries that carry a 'defined' position, but otherwise only consist of 'mixed' positions, are required for every position to be decoded. They are then subjected to biological screening, and the most active groups are selected at every position; the most active sequence is determined by this in the ideal case.

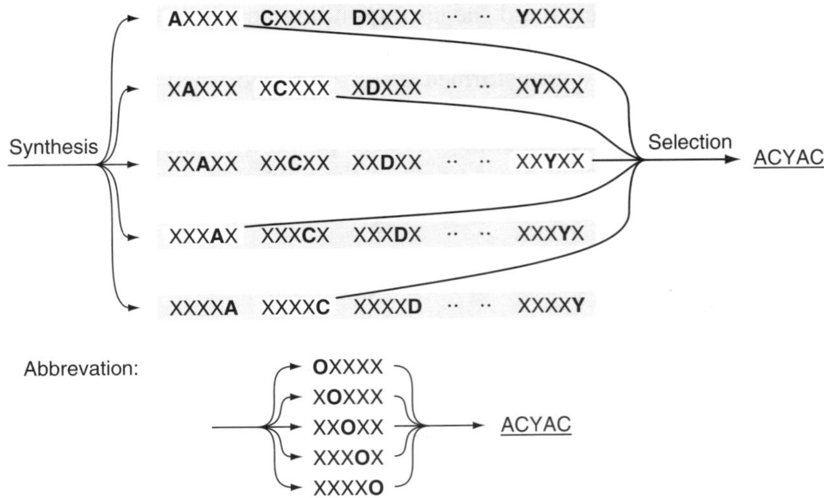

Figure 4.24 Positional scanning with the example of the deconvolution of a pentapeptide.

Sublibraries of the pentapeptide library that carry a 'defined' amino acid at the corresponding position, but otherwise only consist of 'mixed' positions, are required for every amino acid position. Five ('defined' positions) $\times$ 20 (natural amino acids at the 'defined' position) $=$ 100 sublibraries have to accordingly be synthesized for the positional scanning of a pentapeptide. They are then subjected to biological screening, and the most active amino acids at every position are selected. Ideally, the most active sequence is determined here, which still has to be verified by the synthesis and biological testing of the corresponding individual peptide. The two or three most active amino acids at every sequence position are normally selected, however, so a two-digit number of individual peptides have to be subsequently synthesized and screened.

A complementary deconvolution strategy called deletion synthesis deconvolution was published recently [Boger *et al.*, 1998].

Orthogonal Deconvolution

The term 'orthogonal' deconvolution goes back to Déprez *et al.* [1995]; similar approaches were simultaneously introduced by various other groups [Smith *et al.*, 1994; Pirrung and Chen, 1995; Bray, 1996]. The terms 'indexed combinatorial libraries' [Pirrung and Chen, 1995] and 'transformed group libraries' [Bray, 1996] are therefore also used in connection with this. The principle procedure is to be explained in detail next with the example of the deconvolution of a hypothetical dimer library that includes 35 members and that arose through the reaction of five electrophiles (acid chlorides) with seven nucleophiles (primary amines) (Figure 4.25).

Nucleophile sublibraries

Electrophile sublibraries	N_1	N_2	N_3	N_4	N_5	N_6	N_7
E_1	e_1n_1	e_1n_2	...	...	...	...	e_1n_7
E_2	e_2n_1						
E_3	...						
E_4	...						
E_5	e_5n_1						e_5n_7

Figure 4.25 Orthogonal deconvolution of a hypothetical dimer library that includes 35 members and that arises through the reaction of five electrophiles with seven nucleophiles (see text for an explanation).

The starting point of the deconvolution is the synthesis of five electrophile sublibraries $E_1 - E_5$ by choosing an electrophile e_x in each case and reacting it with a mixture of the seven nucleophiles $n_1 - n_7$. Analogously to this, one synthesizes seven nucleophile sublibraries $N_1 - N_7$ by choosing a nucleophile n_x in each case and reacting it with a mixture of the five electrophiles $e_1 - e_5$. The five electrophile sublibraries therefore contain the same 35 product dimers e_xn_x (acid amides) as the seven nucleophile sublibraries, but in a more or less 'orthogonal' distribution. The most active dimer can be subsequently decoded by screening the 12 sublibraries (Figure 4.26). The 'most interesting' product is namely potentially located in the most active electrophile library as well as in the most active nucleophile library, so the intersection from these two sublibraries ideally represents the product that is being looked for. However, this has to still be verified by synthesis and biological testing of the corresponding individual compound (cf. Smith *et al.*, [1994]; and Pirrung and Chen [1995]).

This approach, with appropriate modification, can also be applied to trimer and oligomer libraries [Déprez *et al.*, 1995; Bray, 1996].

Recursive Deconvolution

The method of recursive deconvolution [Erb *et al.*, 1994] is based on the divide, couple, and recombine technique (see earlier section in this Chapter), but requires a slight change. Before the polymers are recombined after the separate coupling of the amino acids, a small part of every intermediate has to be set aside, cataloged and stored (Figure 4.27). These intermediates are for recursive

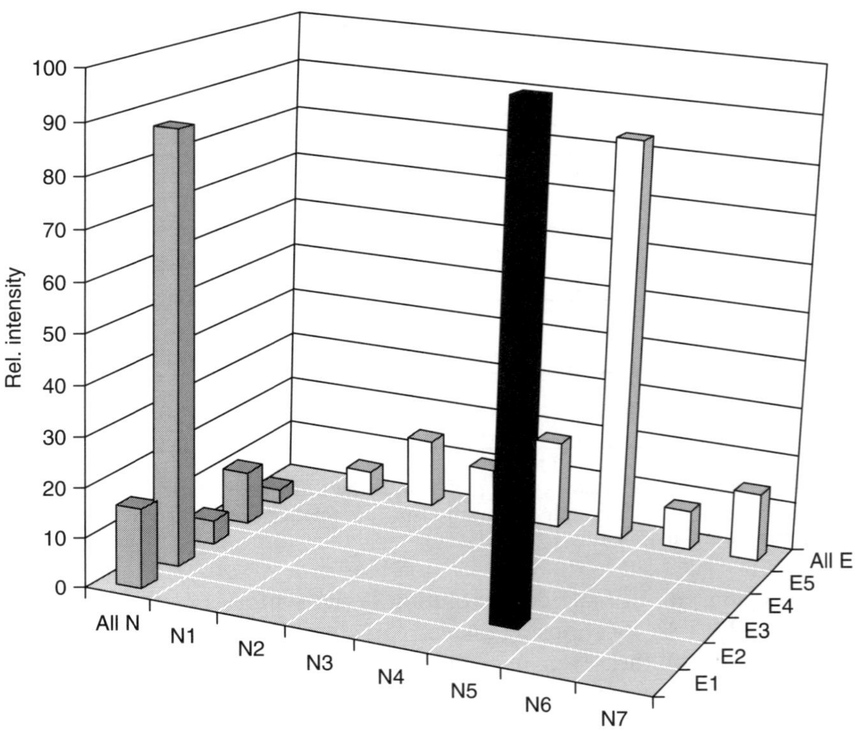

Figure 4.26 Orthogonal deconvolution of a hypothetical dimer library: result of the screening. White: seven nucleophile sublibraries; gray: five electrophile sublibraries; black: 'the most active' individual compound at the interface of the most active sublibraries (see text for an explanation).

determination of the active sequence in the further process, which is to be explained in detail with the example of recursive deconvolution of a tripeptide (Figure 4.28).

After synthesis of the sublibraries of the type OXX ('defined' amino acid at the *N*-terminus, all of the remaining positions are 'mixed' positions), they are – either on the polymer or in solution – screened in a suitable assay system. The sublibrary with the best results is selected, and the corresponding 'defined' amino acid is coupled to the stored intermediate sublibraries of the type OX. The tripeptide sublibraries that arise are screened again; the two *N*-terminal amino acids with the best properties can be decoded because of this. They are then coupled to the stored intermediate sublibraries of the type O, so the active tripeptide can be completely determined in a last screening.

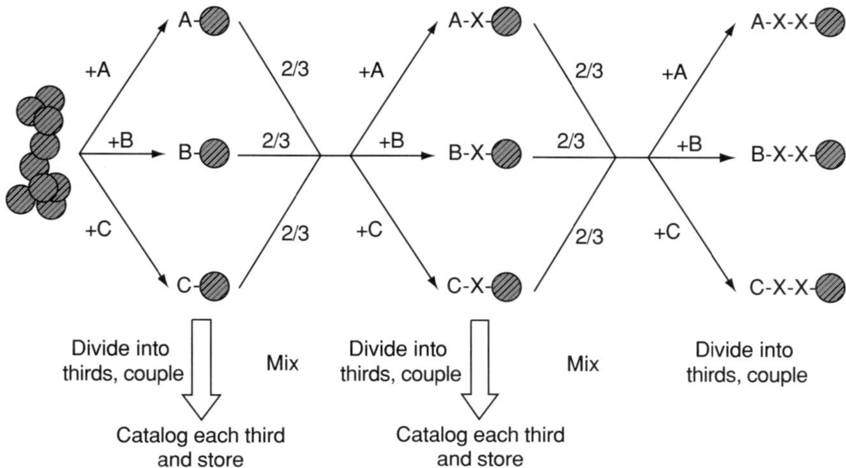

Figure 4.27 Generation of a library, the 'most interesting' compound of which is to be decoded with the aid of recursive deconvolution [Erb *et al.* 1994]. As a distinction to the divide, couple, and recombine technique, a small part of each intermediate has to be set aside, cataloged and stored.

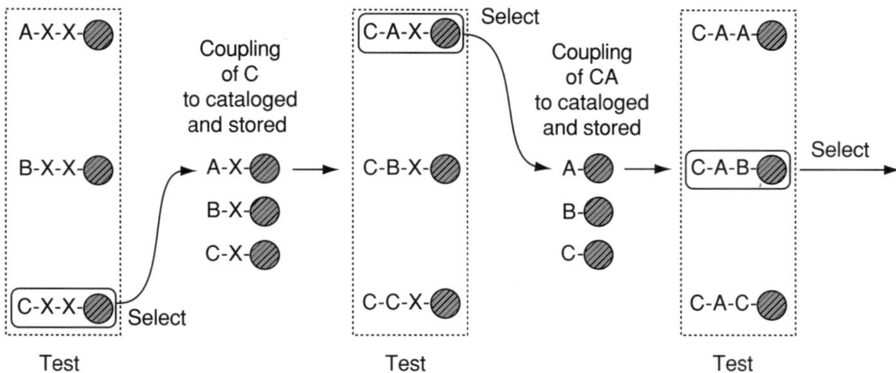

Figure 4.28 Recursive deconvolution according to Erb *et al.* [1994]. The 'defined' amino acid that provides the 'most interesting' results within the framework of screening of the OXX sublibraries is coupled to the intermediate sublibraries of the type OX that were stored during the synthesis (see Figure 4.27). The process is carried on according to this scheme until the 'most interesting' sequence is completely deduced.

REFERENCES

Barnes, C. and Balasubramanian, S. (2000). Recent developments in the encoding and deconvolution of combinatorial libraries, *Curr. Opin. Chem. Biol.* **4**, 346–350.

Boger, D. L., Chai, W. and Jin, Q. (1998). Multistep convergent solution-phase combinatorial synthesis and deletion synthesis deconvolution, *J. Am. Chem. Soc.* **120**, 7220–7225.

Borchardt, A. and Still, W. C. (1994). Synthetic receptor for internal residues of a peptide chain. Highly selective binding of (L)X–(L)Pro–(L)X tripeptides, *J. Am. Chem. Soc.* **116**, 7467–7468.

Bray, A. M. (1996). The transformed group library method: a new library design and mixture decode strategy. In: Smith, J. and Rivier, J. E. (eds), *Peptides: Chemistry, Structure and Biology (Proceedings of the 14th American Peptide Symposium)*, Mayflower Scientific Ltd, Kingswinford, UK, pp. 290–292.

Brenner, S. and Lerner, R. A. (1992). Encoded combinatorial chemistry, *Proc. Natl. Acad. Sci. USA*, **89**, 5381–5383.

Câmpian, E. Sebestyén, F. and Furka Á (1994). Colored and fluorescent solid supports, 469–472. In: *Innovation and Perspectives in Solid Phase Synthesis: Peptides, Proteins and Nucleic Acids* (Edited by R. Epton), Mayflower, Birmingham, UK.

Carell, T. Wintner, E. A. and Rebek, J. (1994). A solution-phase screening-procedure for the isolation of active compounds from a library of molecules, *Angew. Chem. Int. Ed. Engl.* **33**, 2061–2064.

Chen, C. T. Wagner, H. and Still, W. C. (1998). Fluorescent, sequence-selective peptide detection by synthetic small molecules, *Science*, **279**, 851–853.

Chu, Y. H. Avila, L. Z. Biebuyck, H. A. and Whitesides, G. M. (1993). Using affinity capillary electrophoresis to identify the peptide in a peptide library that binds most tightly to vancomycin, *J. Org. Chem.* **58**, 648–652.

Chu, Y. H. Dunayevskiy, Y. M. Kirby, D. P. Vouros P. and Karger, B. L. (1996). Affinity capillary electrophoresis mass-spectrometry for screening combinatorial libraries, *J. Am. Chem. Soc.* **118**, 7827–7835.

Déprez, B. Williard, X. Bourel, L. Coste, H. Hyafil, F. and Tartar A. (1995). Orthogonal combinatorial chemical libraries, *J. Am. Chem. Soc.* **117**, 5405–5406.

Ecker, D. J. Vickers, T. A. Hanecak, R. Driver, V. and Anderson, K. (1993). Rational screening of oligonucleotide combinatorial libraries for drug discovery, *Nucleic Acids Res.* **21**, 1853–1856.

Eckes P. (1994). Binary encoding of compound libraries, *Angew. Chem. Int. Ed. Engl.* **33**, 1573–1575.

Egner, B. J. Rana, S. Smith, H. Bouloc, N. Frey, J. Brocklesby W. S. and Bradley, M. (1997). Tagging in combinatorial chemistry: the use of colored and fluorescent beads, *Chem. Commun.* **8**, 735–736.

Ellington, A. D. and Szostak, J. W. (1990). *In vitro* selection of RNA molecules that bind specific ligands, *Nature*, **346**, 818–822.

Erb, E. Janda, K. D. and Brenner, S. (1994). Recursive deconvolution of combinatorial chemical libraries, *Proc. Natl. Acad. Sci. USA*, **91**, 11 422–11 426.

Furka, Á. Sebestyén, F. Asgedom, M. and Dibó, G. (1988a). *14th Int. Congr. Biochem.* Prague, Czechoslovakia, p. 47.

Furka, Á. Sebestyén, F. Asgedom, M. and Dibó, G. (1988b). *10th Int. Symp. Med. Chem.* Budapest, Hungary, p. 168.

Furka, Á. Sebestyén, F. Asgedom, M. and Dibó, G. (1991). General method for rapid synthesis of multicomponent peptide mixtures, *Int. J. Pept. Protein Res.* **37**, 487–493.

Geysen, H. M. Rodda S. J. and Mason, T. J. (1986). *A priori* delineation of a peptide which mimics a discontinuous antigenic determinant, *Mol. Immunol.* **23**, 709–715.

Geysen, H. M. Wagner, C. D. Bodnar, W. M. Markworth, C. J. Parke, G. J. Schoenen, F. J. Wagner, D. S. and Kinder, D. S. (1996). Isotope or mass encoding of combinatorial libraries, *Chem. Biol.* **3**, 679–688.

Gordeev, M. F. Patel, D. V. and Gordon, E. M. (1996). Approaches to combinatorial synthesis of heterocycles – a solid-phase synthesis of 1,4-dihydropyridines, *J. Org. Chem.* **61**, 924–928.

Guiles, J. W. Lanter, C. L. and Rivero, R. A. (1998). A visual process for mix and sort combinatorial chemistry, *Angew. Chem. Int. Ed. Engl.* **37**, 926–928.

Houghten, R. A. Pinilla, C. Appel, J. R. Blondelle, S. E. Dooley, C. T. Eichler, J. Nefzi, A. and Ostresh, J. M. (1999). Mixture-based synthetic combinatorial libraries, *J. Med. Chem.* **42**, 3743–3778.

Houghten, R. A. Pinilla, C. Blondelle, S. E. Appel, Dooley, J. R. C. T. and Cuervo, J. H. (1991). Generation and use of synthetic peptide combinatorial libraries for basic research and drug discovery, *Nature*, **354**, 84–86.

Ivanetich, K. M. and Santi, D. V. (1996). Preparation of equimolar mixtures of peptides by adjustment of activated amino acid concentrations, 247–260. In: Abelson, J. N. (ed.), *Combinational Chemistry* vol. 267, Academic Press, San Diego,

Jacobs, J. W. and Ni, Z.-J. (1998). Encoded combinatorial chemistry, 271–290. In: *Combinatorial Chemistry and Molecular Diversity in Drug Discovery* (Edited by E. M. Gordon and J. F. J. Kerwin), John Wiley & Sons, New York.

Janda, K. D. (1994). Tagged versus untagged libraries: methods for the generation and screening of combinatorial chemical libraries. *Proc. Natl. Acad. Sci. USA*, **91**, 10 779–10 785.

Jayawickreme, C. K. Graminski, G. F. Quillan J. M. and Lerner, M. R. (1994). Creation and functional screening of a multi-use peptide library. *Proc. Natl. Acad. Sci. USA*, **91**, 1614–1618.

Jung G. and Beck-Sickinger, A. G. (1992). Multiple peptide-synthesis methods and their applications. *Angew. Chem. Int. Ed. Engl.* **31**, 367–383.

Kaur, S. McGuire, L. Tang, D. Dollinger G. and Huebner, V. (1997). Affinity selection and mass spectrometry-based strategies to identify lead compounds in combinatorial libraries, *J. Protein Chem.* **16**, 505–511.

Kelly, M. A. Liang, H. Sytwu, I. I. Vlattas, I. Lyons, N. L. Bowen B. R. and Wennogle, L. P. (1996). Characterization of SH2 – ligand interactions via library affinity selection with mass spectrometric detection, *Biochemistry*, **35**, 11 747–11 755.

Kerr, J. M. Banville, S. C. and Zuckermann, R. N. (1993). Encoded combinatorial peptide libraries containing nonnatural amino-acids, *J. Am. Chem. Soc.* **115**, 2529–2531.

Lam, K. S. Lebl, M. and Krchnak, V. (1997). The 'one-bead-one-compound' combinatorial library method, *Chem. Rev.* **97**, 411–448.

Lam, K. S. Salmon, S. E. Hersh, E. M. Hruby, V. J. Kazmierski, W. M. and Knapp, R. J. (1991). A new type of synthetic peptide library for identifying ligand-binding activity, *Nature*, **354**, 82–84.

Lew, A. M. and Kemp, D. J. (1989). Isolating DNA segments from cloned libraries without screening by affinity selection of PCR products, *Nucleic Acids Res.* **17**, 5859–5860.

Moran, E. J. Sarshar, S. Cargill, J. F. Shahbaz, M. M. Lio, A. Mjalli, M. M. and Armstrong, R. W. (1995). Radio-frequency tag encoded combinatorial library method for the discovery of tripeptide-substituted cinnamic acid inhibitors of the protein-tyrosine-phosphatase Ptp1B, *J. Am. Chem. Soc.* **117**, 10 787–10 788.

Needels, M. C. Jones, D. G. Tate, E. H. Heinkel, G. L. Kochersperger, L. M. Dower, W. J. Barrett R. W. and Gallop, M. A. (1993). Generation and screening of an oligonucleotide-encoded synthetic peptide library, *Proc. Natl. Acad. Sci. USA*, **90**, 10 700–10 704.

Nestler, H. P. Wennemers, H. Sherlock, R. and Dong, D. Y. (1996). Microautoradiographic identification of receptor–ligand interactions in bead-supported combinatorial libraries, *Bioorg. Med. Chem. Lett.* **6**, 1327–1330.

Ni, Z. J. Maclean, D. Holmes, C. P. Murphy, M. M. Ruhland, B. Jacobs, J. W. Gordon E. M. and Gallop, M. A. (1996). Versatile approach to encoding combinatorial organic syntheses using chemically robust secondary amine tags, *J. Med. Chem.* **39**, 1601–1608.

Nicolaou, K. C. Xiao, X. Y. Parandoosh, Z. Senyei A. and Nova, M. P. (1995). Radiofrequency encoded combinatorial chemistry, *Angew. Chem. Int. Ed. Engl.* **34**, 2289–2291.

Nielsen, J. Brenner, S. and Janda, K. D. (1993). Synthetic methods for the implementation of encoded combinatorial chemistry, *J. Am. Chem. Soc.* **115**, 9812–9813.

Nikolaiev, V. Stierandová, A. Krchnák, V. Seligmann, B. Lam, K. S. Salmon, S. E. and Lebl, M. (1993). Peptide-encoding for structure determination of nonsequenceable polymers within libraries synthesized and tested on solid-phase supports, *Pept. Res.* **6**, 161–170.

Ohlmeyer, M. H. Swanson, R. N. Dillard, L. W. Reader, J. C. Asouline, G. Kobayashi, R. Wigler M. and Still, W. C. (1993). Complex synthetic chemical libraries indexed with molecular tags, *Proc. Natl. Acad. Sci. USA*, **90**, 10 922–10 926.

Pinilla, C. Appel, J. R. Blanc, P. and Houghten, R. A. (1992). Rapid identification of high affinity peptide ligands using positional scanning synthetic peptide combinatorial libraries, *BioTechniques*, **13**, 901–905.

Pirrung, M. C. and Chen, J. (1995). Preparation and screening against acetylcholinesterase of a nonpeptide indexed combinatorial library, *J. Am. Chem. Soc.* **117**, 1240–1245.

Salmon, S. E. Lam, K. S. Lebl, M. Kandola, A. Khattri, P. S. Wade, S. Patek, M. Kocis, P. Krchnák, V. Thorpe, D. and Felder, S. (1993). Discovery of biologically active peptides in random libraries: solution-phase testing after staged orthogonal release from resin beads, *Proc. Natl. Acad. Sci. USA*, **90**, 11 708–11 712.

Salmon, S. E. Liu-Stevens, R. H. Zhao, Y. Lebl, M. Krchnák, V. Wertman, K. Sepetov, N. F. and Lam, K. S. (1996). High-volume cellular screening for anticancer agents with combinatorial chemical libraries: a new methodology, *Mol. Divers.* **2**, 57–63.

Smith, P. W. Lai, J. Y. Q. Whittington, A. R. Cox, B. Houston, J. G. Stylli, C. H. Banks, M. N. and Tiller, P. R. (1994). Synthesis and biological evaluation of a library containing potentially 1600 amides/esters. A strategy for rapid compound generation and screening, *Bioorg. Med. Chem. Lett.* **4**, 2821–2824.

Songyang, Z. Shoelson, S. E. Chaudhuri, M. Gish, G. Pawson, T. Haser, W. G. King, F. Roberts, T. Ratnofsky, S. Lechleider, R. J. Neel, B. G. Birge, R. B. Fajardo, J. E. Chou, M. M. Hanafusa, H. Schaffhausen, B. and Cantley, L. C. (1993). SH2 domains recognize specific phosphopeptide sequences, *Cell*, **72**, 767–778.

Stefanovic, S. Wiesmueller, K. H. Metzger, J. W. Beck-Sickinger, A. G. and Jung, G. (1993). Natural and synthetic peptide pools: characterization by sequencing and electrospray mass spectrometry, *Bioorg. Med. Chem. Lett.* **3**, 431–436.

Tan, D. S. and Burbaum, J. J. (2000). Ligand discovery using encoded combinatorial libraries, *Curr. Opin. Drug Discovery Dev.* **3**, 439–453.

Tuerk, C. and Gold, L. (1990). Systematic evolution of ligands by exponential enrichment: RNA ligands to bacteriophage T4 DNA polymerase, *Science*, **249**, 505–510.

Xiao, X.-Y. and Nova, M. P. (1997). Radiofrequency encoding and additional techniques for the structure elucidation of synthetic combinatorial libraries, *Comb. Chem.* 135–152.

Xiao, X. Y. Zhao, C. Potash, H. and Nova, M. P. (1997). Combinatorial chemistry with laser optical encoding, *Angew. Chem. Int. Ed. Engl.* **36**, 780–782.

Zhao, Z. G. and Lam, K. S. (1997). Synthetic peptide libraries, 192–209. In: *Annual Reports in Combinational Chemistry and Molecular Diversity* (Edited by W. H. Moos, M. R. Pavia, A. D. Ellington and B. K. Kay), ESCOM, Leiden, the Netherlands.

Zuckermann, R. N. Kerr, J. M. Siani, M. A. Banville, S. C. and Santi, D. V. (1992). Identification of highest-affinity ligands by affinity selection from equimolar peptide mixtures generated by robotic synthesis, *Proc. Natl. Acad. Sci. USA*, **89**, 4505–4509.

5 Parallel Syntheses and Automation

5.1 INTRODUCTION

Whereas development of combinatorial libraries progressed with a great deal of attention paid to it in the past few years, a second development was taking place in silence: Optimization and further development of parallel synthesis, mainly in the chemical laboratories of large pharmaceutical companies. The objective here was to use the advantages of combinatorial libraries – the large number of compounds – but to bypass the disadvantages, for instance problematic identification of the active compounds [Hardin and Smietana, 1996; Bondy, 1998].

In principle, individual compounds are synthesized in parallel synthesis. In contrast to classical chemistry, variable positions (x-, y-, z-positions) are defined, however, as in the case of libraries. A decision is therefore made, as an example, that a carboxyl group is to be reacted with 10 different alcohols in parallel to form 10 different esters (x-position). One proceeds in an analogous fashion with the other variable positions. After this, every x is combined synthetically with every y and every z in a parallel fashion; a so-called array, also called a product matrix, arises because of this [Burbaum and Sigal, 1997; Hill, 1998]. The tests of the individual compounds then follow after the synthesis of the entire array.

It is not possible to synthesize millions of different compounds with this approach as far as the concept is concerned, but the array synthesis is used less frequently for finding new lead compounds than for the optimization of lead compounds that are already known. If a compound with pharmaceutically interesting properties has been found by the screening of either a combinatorial library, or natural substance pools or the synthetic compound pool of a company, it has to generally be optimized with regard to various parameters. In this context, it is of particular interest to increase the selectivity, the activity, the efficiency, and the bioavailability and to lower the toxicity [Kubinyi, 1998]. Parallel synthesis methods are especially suitable for these optimizations.

In classical medicinal chemistry one to 10 compounds were manually synthesized one after the other and it depended mainly on experience, intuition, or serendipity as to which modifications could be successful under certain circumstances. In combinatorial approaches, however, this influences the synthetic strategy only sporadically. The possible variable positions of the lead compound are first determined and varied independently of one another. A great deal of importance is attached to the development of a suitable synthesis strategy in this context. Which variations are possible at which synthesis step are established based on this. Three to five variables are customary in the process, for example functional groups, different chain lengths and the introduction of heteroatoms into the core structure (exchange of CH_2 with O, NH, or S). If, for example, two different functional groups are modified with 10 reaction partners each and, in addition, three different chain lengths and the use of a further heteroatom are taken into consideration, a four-dimensional matrix consisting of $10^2 \times 3 \times 2 = 600$ individual compounds arises.

It is self-explanatory that their synthesis would occupy several chemists for years with manual approaches, which is why the parallel, combinatorial synthesis of arrays only prevailed alongside the development of synthesis robots. Outstanding synthesis specialists who must be familiar with both solid phase synthesis strategies as well as synthesis strategies in solution with all of their advantages and disadvantages, are required for development and adaptation of the final synthesis strategy. This strategy decides the success and failure of the entire approach. A few manual test syntheses are frequently made in advance to evaluate the strategy before an attempt is made to transfer the synthesis to robots. The subsequent synthesis of the array then runs on a far smaller scale than is the case with classic syntheses [Harrison, 1998].

5.2 AUTOMATION OF THE SYNTHESIS

A whole series of synthesis robots have come on the market in the past few years; however, the number of concepts is limited. A few typical representatives will be introduced below and, using them as an example, the essential basics will be explained. A distinction is made in principle between:

- semiautomated synthesizers which allow manual steps in part;
- fully automated synthesizers which independently carry out the synthesis;
- fully automated platforms or workstations which handle the synthesis and all purification steps, possibly even including analysis and screening.

In addition to the devices presented in this book, there is a multitude of other approaches employed by various companies which are significantly different in their details under certain circumstances, but usually have a very similar structure as far as the principle is concerned. There are quite a number of recent reviews about these approaches [Cargill and Lebl, 1997; Needels and Sugarman, 1998; Antonenko, 1999; Hird, 1999; Powers and Coffen, 1999; Thiericke *et al.*, 1999].

5.2.1 FROM SEMIAUTOMATED TO FULLY AUTOMATED SYNTHESIZER

It usually is advisable not to use a fully automated synthesizer for the development of new syntheses or to explore suitable strategies. A better alternative would be a machine that, for example, automates the washing steps, but with which the addition of reagents as well as adjustment of the reaction conditions can be individually variably, and even manually controlled.

Device APOS 1200 (Rapp Polymere, Tübingen, Germany) allows the parallel solid phase synthesis of 12 products per reaction block under an inert gas atmosphere. Each vessel can be addressed individually including reaction time and temperature (−60°C to +170 °C) (Figure 5.1). Each vessel is equipped with a glass frit at the bottom and with a reflux condenser at the top. The addition of reagents is done manually with the aid of a syringe or – in the fully automated version – with the aid of a pipetting robot (Figure 5.2). The reaction block can take two positions in the device in principle: In the first position, an inert gas flow can be guided from below through the glass frit into the reaction vessel for thorough mixing of the reaction partners and drained with the aid of a vacuum exhauster mounted on the top of the condensers. In the second position, the reaction or washing solutions of the 12 vessels can be drained through the frits to a waste bottle. If the products are to be separately collected after cleavage

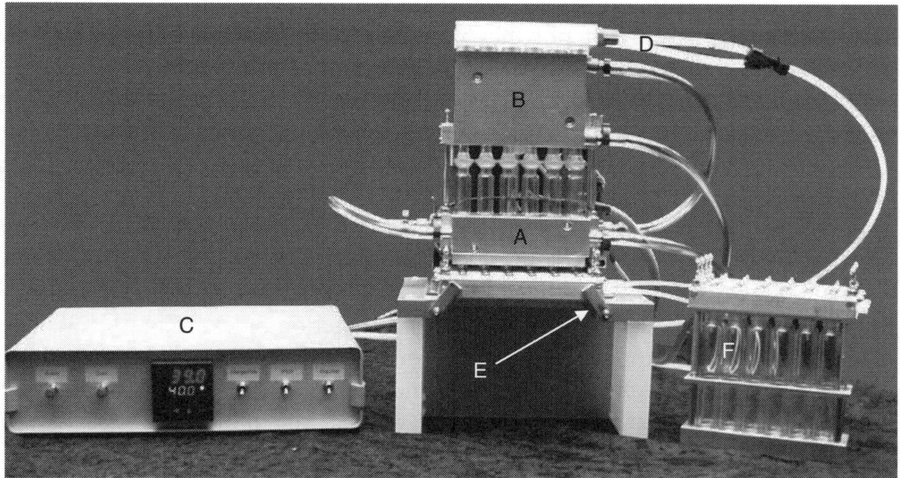

Figure 5.1 APOS 1200 semiautomatic synthesis block. A: Twelve synthesis vessels. B: Reflux condenser set on top. C: Control unit for temperature, gas, and vacuum. D: Valve system for gas/vacuum connection. E: Transfer unit, allows to jointly drain the reaction solutions or to separately collect the products after cleavage. F: Unit for collecting the products: The reaction and collection vessels have to be transferred into the transfer unit (arrow) for this. (The illustration was kindly made available by Rapp Polymere, Tübingen, Germany.)

Figure 5.2 APOS 1200 synthesis unit. A: Expanded into a fully automated system by integration into a pipetting robot (Tecan MSP 75/1 or MSP 75/2). The liquid is added through the pipetting needle (B) of the arm (C), which is freely movable in the *x*-, *y*-, and *z*-directions. The robot is controlled with the aid of a PC (D). (The illustration was kindly made available by Rapp Polymere, Tübingen, Germany.)

from the resin, position 2 can likewise be used for this, but a different bottom plate that permits separate collection of the product solutions has to be used. The scale size is 200 mg of polymer with 2 ml of solvent per reaction vessel, which corresponds to approximately 0.1–0.2 mmol of product.

Device SAS (semiautomated system) of MultiSynTech (Witten, Germany) for which the reaction block and configuration scheme are schematically depicted in Figures 5.3 and 5.4, respectively, is likewise designed for solid phase organic syntheses in a broad temperature range (−30 °C to +150 °C) under an inert gas atmosphere. In contrast to the device that was described previously, various reaction blocks are offered (from 40 vessels having a size of 10 ml up to 96 vessels having a size of 2 ml) that permit parallel synthesis on a larger scale. A three-way valve plate serves as the base for the reaction blocks (Figure 5.3). The reaction vessels are supplied with frit bottoms. Thus, solvent can be sucked off when a vacuum is applied to the plate and inert gas can flow in the reverse direction. The entire reaction block, as well as the individual vessels, are covered from above with septa. The novel aspect of the reaction blocks, however, is the elegant solution for thorough mixing of the individual reaction partners: Electric coils are wrapped around each reaction vessel, which generate a magnetic field that drives the stirring bars placed in the vessels in a levitated way, a few millimeters above the frit bottoms. Due to this technique the resin is not crunched between the stirring bar and the frit during rotation, and gentle mixing is made possible.

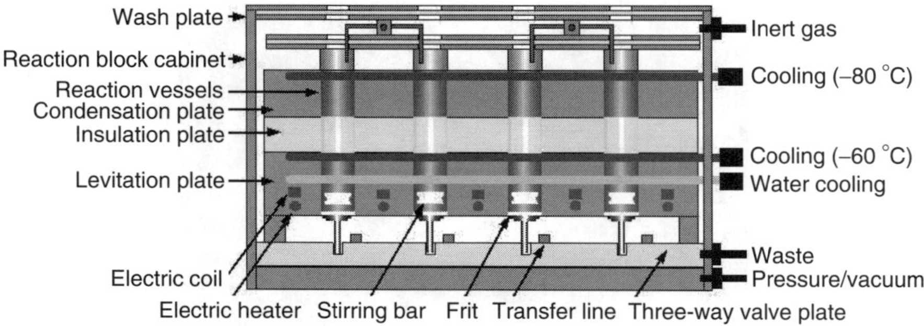

Figure 5.3 Schematic diagram of the reaction block of the semiautomatic SAS synthesizer. (The illustration was kindly made available by MultiSynTech, Witten, Germany.)

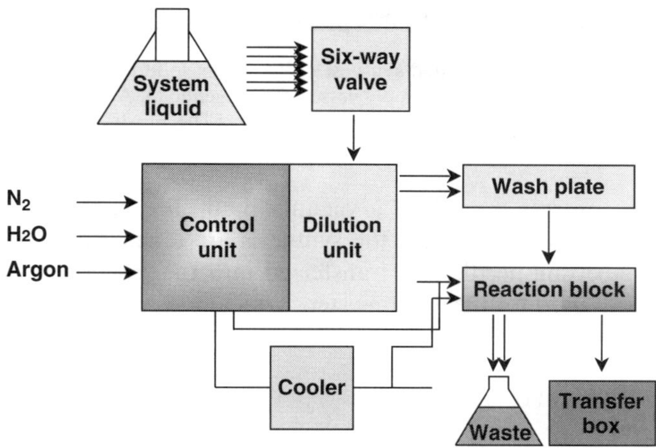

Figure 5.4 Configuration diagram of the semiautomatic SAS synthesizer. (The illustration was kindly made available by MultiSynTech, Witten, Germany.)

5.2.2 A FULLY AUTOMATED DEVICE FOR SOLID PHASE SYNTHESIS

The fully automated synthesizer 'Syro' was developed by MultiSynTech for parallel, solid phase synthesis of organic compounds through the combination of a pipetting robot from Tecan (Männedorf, Switzerland) with the reaction block design described at the end of the previous section (Figure 5.5). The filling of the reaction vessels which are fixed in the reaction block is done here by the pipetting robot which is controlled by a personal computer (PC) with corresponding software. Depending on the model, the reaction block can hold between 40 vessels (10 ml each) and 96 vessels (2 ml each), as in the case of the SAS device.

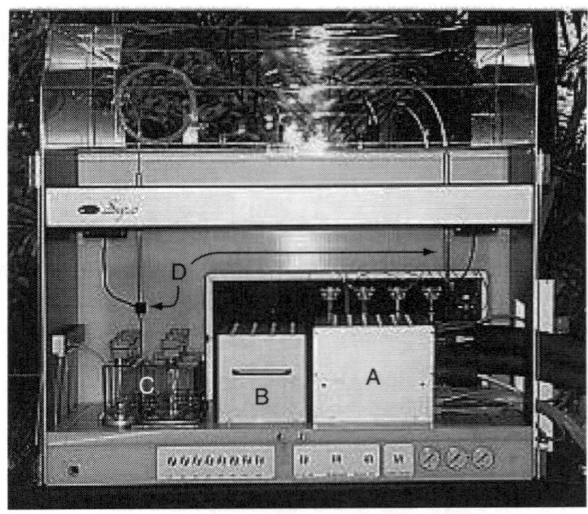

Figure 5.5 Multiple, fully automated synthesizer 'Syro' for solid phase synthesis of organic molecules. A: Reaction block. B, C: Supply vessels for reagents. D: Arm with pipetting needle which is freely movable in the *x*-, *y*-, and *z*-directions. (The illustration was kindly made available by MultiSynTech, Witten, Germany.)

The PC also controls the use of a vacuum pump for emptying the reaction vessel. The reagents required for the synthesis are taken from special supply vessels by a pipetting needle and transferred into the reaction vessels (Figure 5.5). The solvent used for the washing steps typically serves as the system liquid.

5.2.3 A FULLY AUTOMATED DEVICE FOR SPOT SYNTHESIS

A method that has especially come into prominence in the area of diagnostics in the past few years is spot synthesis on cellulose membranes (see Section 2.3.2); testing can take place in a manner analogous to the Western blot. Very small product quantities are required on very small spot areas for this method. This means that the pipetting robots have to pipette the reagents with extreme precision at the required positions and furthermore have to be in a position to pipette a few microliters in a reproducible fashion. Successful use of the ASP222 system by Abimed (Langenfeld, Germany), based on a Gilson robot (Middleton, USA), is documented in Figure 5.6.

5.2.4 A FULLY AUTOMATED SYNTHESIZER, ALSO FOR SYNTHESIS IN SOLUTION

The previously described synthesizers and reaction blocks are exclusively suited to the automation of solid phase syntheses. The expense and effort for

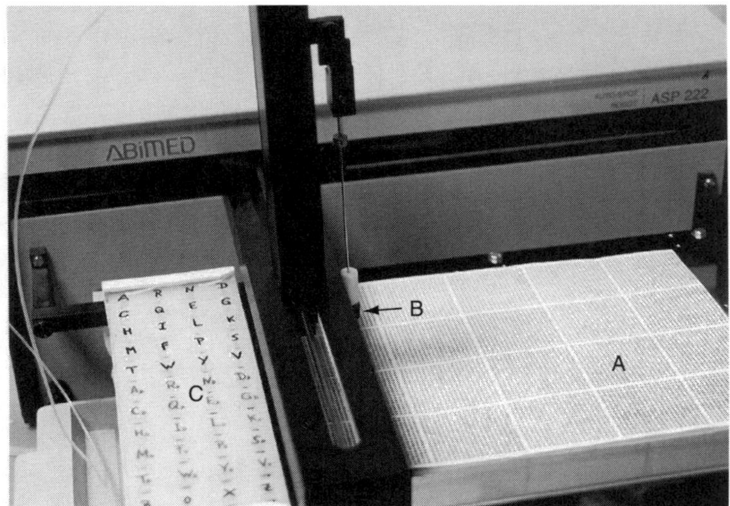

Figure 5.6 ASP222 robot in use for the parallel synthesis of 8000 peptide spots on cellulose. The spots are evenly spaced on a membrane the size of a DIN A4 sheet (A). The specially constructed needle (B) provides the reaction solutions that it previously sucked in from the reservoirs (C) by setting down at defined positions. In addition to the 20 natural amino acids, vessels for 28 nonproteinogenic amino acids are also available, which permits the number of synthesis possibilities to be increased. (The illustration was kindly made available by Jerini, Berlin, Germany.)

development of a suitable synthesis strategy can be considerable for this, however, whereas synthesis in solution without the use of polymers is usually either already known about (if the lead compound comes from the company's own pool, for example) or is quickly accessible. For this reason, the organic synthesis robot presented in Figure 5.7, which is based on the Tecan pipetting robot 'Genesis', can be used with extreme variability: It can be used both for solid phase synthesis, as well as for synthesis in solution. The reaction block consists of 48 round-bottom flasks made of glass that can each hold 10 ml of solution. They are located on top of a reaction block, which allows the pipetting of solutions or flushing of the lines with inert gas; filters have to merely be supplied for use in solid phase synthesis. The block can be shaken and can be transferred as a whole to a cooling or heating module (Figure 5.8). Furthermore, the use of round-bottom flasks allows basic work steps for synthesis in solution, such as extraction for instance (Figure 5.9). The robot uses an eight-needle system for uptake and delivery of the liquids, so the individual work steps can be carried out in a relatively quick way. Detectors that can recognize the various conductivities of the different solvents are located on the tips of the needles. This allows the determination of the phase boundary.

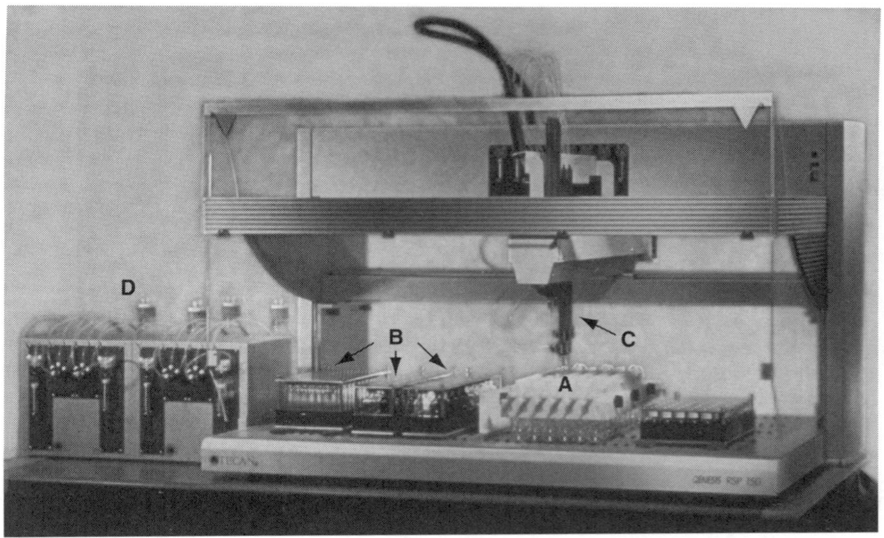

Figure 5.7 Multiple, automated synthesizer which is also suited to carrying out reactions in solution and is based on the pipetting robot 'Genesis'. A: Reaction block (Figure 5.8). B: Reagent supply vessels. C: Multi-channel pipetting needle system. D: Supply vessels of the system liquid (for example solvent). (The illustration was kindly made available by Tecan, Männedorf, Switzerland.)

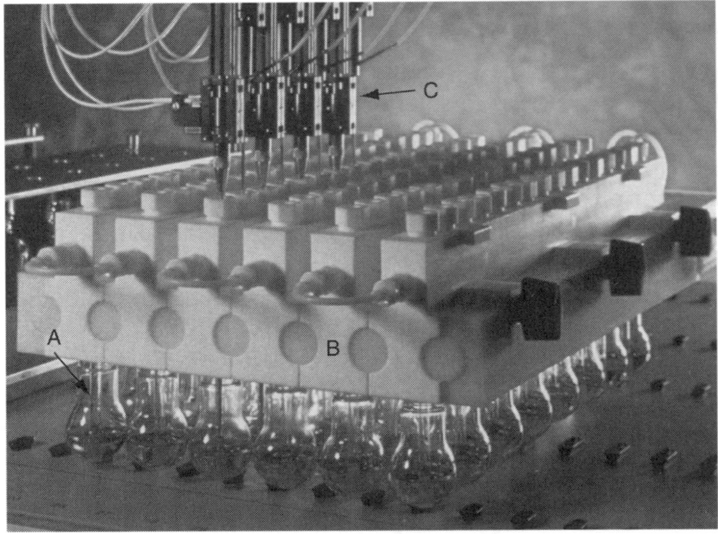

Figure 5.8 Reaction block for the synthesis of 48 reactions in solution or on solid phase. A: Reaction vessel (round-bottom flask made of glass). B: Teflon holder. C: Needle system for supplying and removing liquids. (The illustration was kindly made available by Tecan, Männedorf, Switzerland.)

Figure 5.9 Extraction step in the automated synthesis in solution by a 'Genesis' robot. A: The needle immerses into the reaction vessel, sucks in the lower, colorless phase (the phase borderline is identified with the aid of detectors), and transfers it into the waste vessel. B: The upper, colored phase remains in the flask. (The illustration was kindly made available by Tecan, Männedorf, Switzerland.)

5.2.5 PLATFORMS OR WORKSTATIONS: FULLY INTEGRATED SYSTEMS

The evolution of parallel synthesis runs from the reaction block with several reaction vessels through fully automated synthesizers, up to the so-called synthesis platforms or workstations. These systems are able to perform a multitude of work steps that are necessary within the course of a synthesis and normally consist of various subunits, so-called stations.

Which steps should a system of this type be able to perform? The synthesis preparation requires calculation of the sets, weighing of the reagents, addition of the solvent, and – if solid phase synthesis is involved – weighing of the polymers. The encoding of the reaction vessels is essential here, so that a clear assignment is possible again after the synthesis is carried out or after testing of the activity. Carrying out the reaction itself requires steps such as heating, cooling, shaking and the addition of reagents. Furthermore, one or more stations are required for the purification. This could be filtration, extraction, centrifugation or evaporation. The connection of the individual stations is managed by the transfer of the reaction vessels (individually or in a block) with suitable robot grippers. A synthesis cycle should be able to be programmed, because an entire series of identical reactions is executed repeatedly in the normal case. The system has to be in a position to react in a way that is as variable as possible, because different compounds are to be produced in every vessel and differences inevitably occur because of this in the solubility, reaction speed and selectivity. Adequate washing and filtering steps should be built in case the intermediates or products precipitate.

Nowadays each company often develops its own platform or workstation for its very specific requirements. One of the few systems that is commercially available is 'ARCoSyn' from accelab (Tübingen, Germany) (Figure 5.10). An industrial robot with a precision gripper can move both individual vessels and blocks. Reactions from −30 °C to +120 °C can be completely carried out in an inert gas atmosphere. Depending on the vessel system, work can be done in solution or on the polymeric support. Further stations allow homogenization (dissolution of solid compounds, mixing of various phases, setting of defined concentrations), evaporation (vacuum centrifugation to remove solvents), liquid–liquid extraction, and solid phase extraction. A few of these stations are presented in Figure 5.11.

5.3 MINIATURIZATION OF SYNTHESES AND OF SCREENING

How will development proceed and what is on the horizon in the future? Labaudinière from Rhône-Poulenc Rorer has recently shown how the origin of the various compounds that were investigated in biological tests has changed

Figure 5.10 Overall view of the fully automated platform 'ARCoSyn'. (The illustration was kindly made available by accelab, Tübingen, Germany.)

in the past few years (Figure 5.12) [Labaudinière, 1998]. Over 50% still originated from the internal company pool in 1995; this was down to 19% in 1998. The number of individually synthesized compounds, in contrast, increased in the same period of time from 13 to 56%, whereas the compounds in mixtures reached their maximum in 1997 with 21%. The trend is accordingly heading towards individual compounds that can be produced in parallel and are

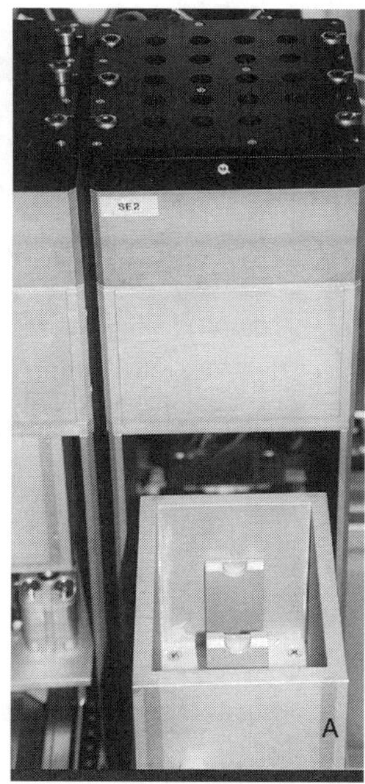

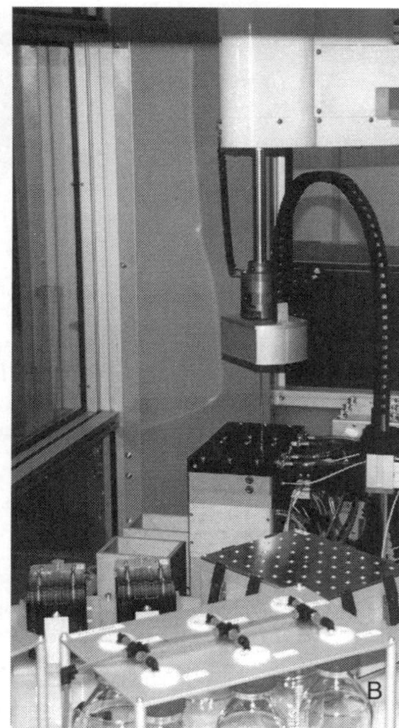

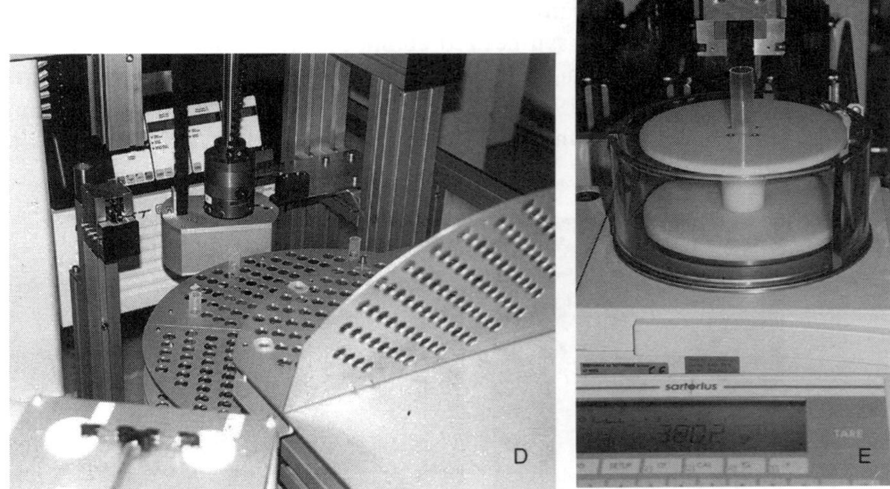

Figure 5.11 Detailed view of individual components of the 'ARCoSyn' synthesis platform. A: Synthesis block. B: Pipetting. C: Centrifugation. D: Transfer into the vacuum centrifuge for removal of solvents or freeze drying. E: Weighing the product yield. (The illustrations were kindly made available by accelab, Tübingen, Germany.)

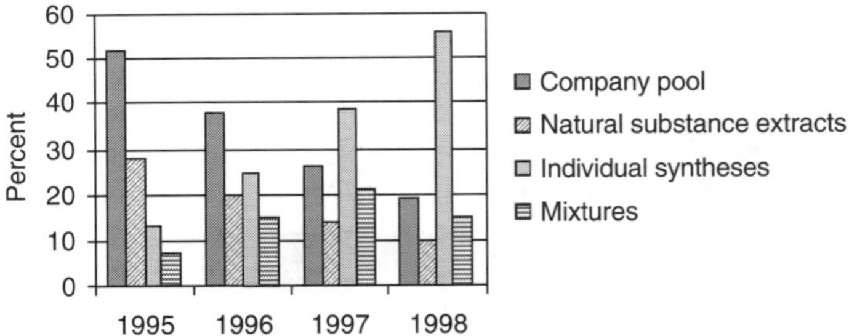

Figure 5.12 Development of composition of the compounds of Rhône-Poulenc Rorer which were investigated in biological tests, arranged according to the origin and synthesis method [Labaudinière, 1998].

amenable to high-throughput screening (HTS) [Burbaum and Sigal, 1997; Finney, 1998; Hill, 1998; Nakayama, 1998].

An attempt is made both during the synthesis and during testing to make do with smaller and smaller quantities [Houston and Banks, 1997]. From the 96-well plates (8 × 12 wells) originally used for screening, it progressed further to 384-well plates in 16 × 24 format, and in the meanwhile 1536-well plates (32 × 48) are already in trials. This requires the pipetting of a few microliters down to nanoliters with a high degree of precision and reproducibility.

The first syntheses on a micro-scale were carried out with the aid of photolithographic processes; up to 40 000 individual compounds were produced on a $1 cm^2$ matrix (see Section 2.3.2). The screening subsequently took place with the aid of a fluorescently labeled antibody. Comparison of the location of the most active areas on the chip with the synthesis scheme subsequently supplied the sequence [Fodor *et al.*, 1991]. Comparable synthesis concepts are being developed and tested in a modified form at present in many laboratories. Micro-channels and micro-reaction vessels are being etched in glass structures or in polymers, for example. Liquids are moved with the aid of electrokinetic pumps or controlled diffusion. It is obvious that these miniaturized reactions are no longer predominantly controlled by chemistry, but also from a series of physical effects [Lemmo *et al.*, 1997].

5.4 THE COMPUTER – MASTER OF DATA?

Bioinformatics is another important area that will gain heavily in importance in future because of automation per se, the growth in information that will come about because of that, and the flood of data in general. It already plays an important role in selection of the compounds to be synthesized, because a compound array should have the greatest possible degree of diversity [Warr, 1997]. This is understood to mean that the structures contained within it should represent the greatest possible degree of chemical and spatial diversity, i.e. as many spatial combinations as possible of hydrogen bond donors, and acceptors, as well as alkaline, acidic, hydrophobic, hydrophilic, aromatic and other properties. Starting with a basic structure, a calculation can be done with the aid of simulation programs to determine which combination of variable residues at certain positions optimally covers the conformation space. A prediction of this type allows to limit the syntheses required or increases the efficiency of the synthesized compounds [Caflisch and Karplus, 1996; Barnard and Downs, 1997; Blaney and Martin, 1997; Mason and Pickett, 1997; Willett, 1997; Kubinyi, 1998].

The next important step is coordination of the data processing. The synthesis and analysis data, the results of biological tests and other chemical and physical parameters of hundreds or thousands of compounds have to be simultaneously recorded. A suitable system that not only collects all of the data, but also processes it, compares it, sorts it, evaluates it, and makes it available to the scientist in a suitable form is indispensable for this. Only in this way can it be ensured that no bottleneck arises due to data-processing that has a later negative effect on the efficiency [Harrison, 1997; Calvert *et al.*, 1999].

REFERENCES

Antonenko, V. V. (1999). Automation in combinatorial chemistry. *Comb. Chem. Technol.* 205–232.

Barnard, J. M. and Downs, G. M. (1997). Computer representation and manipulation of combinatorial libraries. *Perspect. Drug Discov. Des.* **7–8**, 13–30.

Blaney, J. M. and Martin, E. J. (1997). Computational approaches for combinatorial library design and molecular diversity analysis. *Curr. Opin. Chem. Biol.* **1**, 54–59.

Bondy, S. S. (1998). The role of automation in drug discovery. *Curr. Opin. Drug Discov. Dev.* **1**, 116–119.

Burbaum, J. J. and Sigal, N. H. (1997). New technologies for high-throughput screening. *Curr. Opin. Chem. Biol.* **1**, 72–78.

Caflisch, A. and Karplus, M. (1996). Computational combinatorial chemistry for de novo ligand design: review and assessment. *Perspect. Drug Discov. Des.* **3**, 51–84.

Calvert, S. Stewart, F. P. Swarna, K. and Wiseman, J. S. (1999). The use of informatics and automation to remove bottlenecks in drug discovery. *Curr. Opin. Drug Discov. Dev.* **2**, 234–238.

Cargill, J. F. and Lebl, M. (1997). New methods in combinatorial chemistry – robotics and parallel synthesis. *Curr. Opin. Chem. Biol.* **1**, 67–71.

Finney, N. S. (1998). Fluorescence assays for screening combinatorial libraries of drug candidates. *Curr. Opin. Drug Discov. Dev.* **1**, 98–105.

Fodor, S. P. Read, J. L. Pirrung, M. C. Stryer, L. Lu, A. T. and Solas, D. (1991). Light-directed, spatially addressable parallel chemical synthesis. *Science*, **251**, 767–773.

Hardin, J. H. and Smietana, F. R. (1996). Automating combinatorial chemistry: a primer on benchtop robotic systems, *Mol. Divers.* **1**, 270–274.

Harrison, W. (1997). The importance of automated sample management systems in realizing the potential of large compound libraries in drug discovery. *J. Biomol. Screen.* **2**, 203–206.

Harrison, W. (1998). Changes in scale in automated pharmaceutical research. *Drug Discov. Today*, **3**, 343–349.

Hill, D. C. (1998). Trends in development of high-throughput screening technologies for rapid discovery of novel drugs. *Curr. Opin. Drug Discov. Dev.* **1**, 92–97.

Hird, N. W. (1999). Automated synthesis: new tools for the organic chemist. *Drug Discov. Today*, **4**, 265–274.

Houston, J. G. and Banks, M. (1997). The chemical–biological interface: developments in automated and miniaturised screening technology. *Curr. Opin. Biotechnol.* **8**, 734–740.

Kubinyi, H. (1998). Combinatorial and computational approaches in structure-based drug design. *Curr. Opin. Drug Discov. Dev.* **1**, 16–27.

Labaudinière, R. F. (1998). RPRs approach to high-speed parallel synthesis for lead generation. *Drug Discov. Today.* **3**, 511–515.

Lemmo, A. V. Fisher, J. T. Geysen, H. M. and Rose, D. J. (1997). Characterization of an inkjet chemical microdispenser for combinatorial chemistry. *Anal. Chem.* **67**, 543–551.

Mason, J. S. and Pickett, S. D. (1997). Partition-based selection. *Perspect. Drug Discov. Des.* **7–8**, 85–114.

Nakayama, G. R. (1998). Microplate assays for high-throughput screening. *Curr. Opin. Drug Discov. Dev.* **1**, 85–91.

Needels, M. C. and Sugarman, J. (1998). Automation of combinatorial chemistry for large libraries. In: Gordon, E. M. and Kerwin, J. F. J. (eds), *Combinatorial Chemistry and Molecular Diversity in Drug Discovery*, John Wiley & Sons, New York, pp. 339–348.

Powers, D. G. and Coffen, D. L. (1999). Convergent automated parallel synthesis. *Drug Discov. Today*. **4**, 377–383.

Thiericke, R. Grabley, S. and Geschwill, K. (1999). Automation strategies in drug discovery. 56–71. In: Grabley, S. and Thierecke, R. (eds), *Drug Discovery From Nature*, Springer, New York.

Warr, W. A. (1997). Combinatorial chemistry and molecular diversity. An overview. *J. Chem. Inf. Comput. Sci*. **37**, 134–140.

Willett, P. (1997). Computational tools for the analysis of molecular diversity. *Perspect. Drug Discov. Des*. **7–8**, 1–11.

6 Combinatorial Methods With Molecular Biological Techniques

Combinatorial methods are not only successfully implemented with chemical methods, but also with molecular biological methods. Although the result – a mixture of different molecules – is very similar, the procedure is completely different. Only peptides, proteins or nucleic acids can be aquired because the molecules are not obtained chemically, but instead biologically. Various methods have been established in the course of the last 10 years that produce a diversity of molecules either *in vivo*, i.e. in organisms such as bacteria, for example, or *in vitro*, in the test tube with the aid of enzymes or isolated biological components. Either deoxyribonucleic acid (DNA) or ribonucleic acid (RNA) serves as the basis for the production of diversity in both approaches, i.e. mixtures are used instead of individual nucleotides, or variations of the starting DNA/RNA are produced through various molecular biological methods. Depending on the technique, this mixture is then translated into proteins or tested directly as RNA or DNA. Special selection systems are of service here, i.e. methods that filter, or search for the proteins or DNAs/RNAs with the desired properties. The molecules obtained in this fashion are then enriched in a directed way in a further step and subsequently identified.

6.1 PHAGE DISPLAY: VARIATION WITH THE AID OF VIRUSES

Which principle forms the basis? The so-called phage display technology was introduced in 1990 by various groups, in order to obtain protein and peptide libraries [Cwirla *et al.*, 1990; Scott and Smith, 1990]. They used bacteriophages for this, i.e. viruses that infect bacteria. The bacteriophages possess proteins on their surface which tolerate certain changes, for example additional segments in certain sequence areas. With the aid of molecular biological techniques, they now inserted one segment of a mixture of very diverse segments into the corresponding bacteriophage gene instead of a certain additional segment. The viruses infect bacteria; new phages with the modified surface protein are produced because of this. The phages that arise can then be sorted according to their properties, for example the recognition of a certain protein.

Which molecules can these phages additionally carry on their surface? The elements that were first inserted were peptides with a length of 6–15 amino acids which are suitable for characterization of binding sites of polyclonal and monoclonal antibodies, or which recognize receptors, enzymes or even entire viruses. In the meanwhile, however, even large proteins can be inserted into certain specially designed phages [Katz, 1997], which was able to be demonstrated with antibodies, antibody fragments, enzymes and a series of other proteins up to a size of around 80 kDa. Several review articles offer an excellent overview of recent developments in this area [Winter, 1996; Johnsson and Ge, 1999].

6.1.1 THE BACTERIOPHAGES

Viruses that can infect bacteria are called bacteriophages or phages for short [Collins, 1997]. An entire series of so-called filamentous phages are known that can infect gram-negative bacteria such as *Escherichia coli* (*E. coli*, the pet of the molecular biologists). They distinguish themselves by the fact that they contain a single stranded circular DNA that is enclosed in a 7×900 to 7×2000 nm cylinder. Among those that are characterized the best examples are the phages M13, fl and fd; their genomes are completely sequenced and have 98 % sequence identity. They require the so-called F-plasmid (fertility plasmid) of the bacteria for infection and are also called phages of the Ff type. The phage genome encodes for 11 proteins, which is why they and their corresponding genes are designated with the Roman numerals I–XI. The genes additionally get the prefix abbreviation g; the proteins get the corresponding prefix abbreviation p. A series of other ways of designating the proteins are also found. For example, Protein number 8 is also frequently called gpVIII (= gene product VIII), gVIIIp, gp8 or g8p instead of pVIII.

The 11 proteins are responsible for DNA replication (pII and pX), binding to the single stranded DNA (pV) and assembly of the phage (pI, pIV, and pXI) or represent the envelope protein itself (pIII, pVI, pVII, pVIII, pIX) (Figure 6.1).

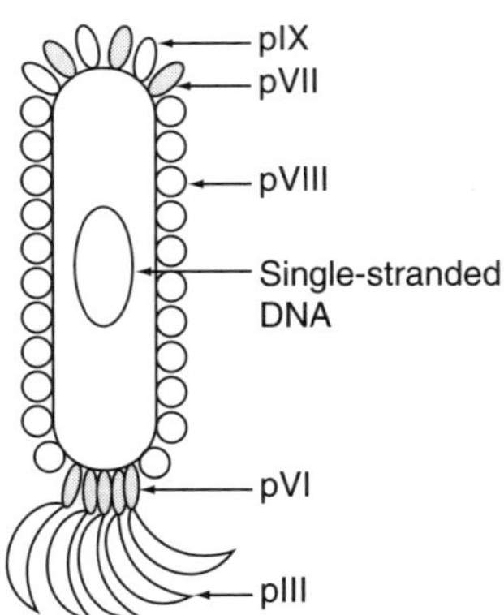

Figure 6.1 Schematic diagram of a phage. The proteins pIII and pVI form the infectious end, the proteins pIX and pVI are on the opposite side. The protein pVIII makes up the main component of the surface.

pVIII with around 2700 copies represents the main part of the phage envelope, consists of 50 amino acids, and has a molecular weight of 5.2 kDa. The positively charged *C*-terminus projects into the cytoplasm when the virus capsid is assembled; the *N*-terminus projects into the periplasm of the bacterial membrane. The four other envelope proteins pIII, pVI, pVII and pIX form the ends of the phage, and each one exists with around five copies. pIII and pVI form the tip, which is responsible for the infection of new bacteria; pVII and pIX form the opposite end (Figures 6.1 and 6.2).

If an Ff phage infects an *E. coli*, it binds with a tip to the F pilus. The tip is a protein tail that is encoded by the F-plasmid and is responsible for the exchange of DNA among the bacteria. The pilus is decomposed, and the phage is brought to the bacterial membrane. The envelope proteins of the phage are integrated into the bacterial membrane and the DNA of the phage is introduced into the cytoplasm. The DNA strand is replicated, and virus production is started in this way with the aid of bacterial enzymes. Viral envelope proteins are formed through the bacterial translation system, built into the bacterial membrane and assembled to new phages together with viral DNA that is stabilized by pV. Bacteria survive the phage infection with a reduction in their generation rate by 50 % and can produce 100–200 phages per phage generation, even 1000 phages in the first generation. Because the size of the phage genome does not influence the replication, Ff phages already found favor early on as tools in

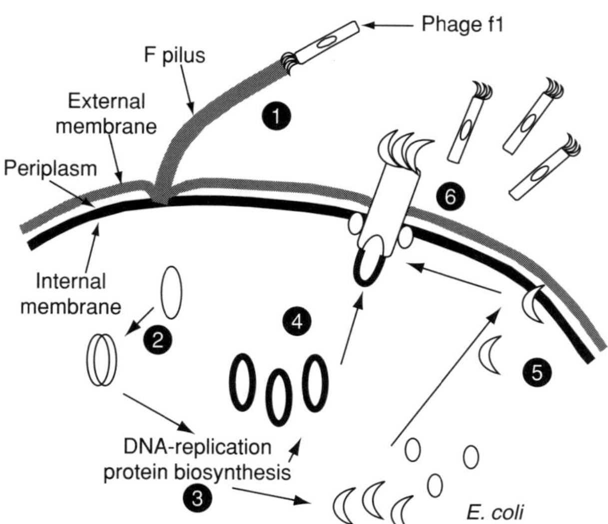

Figure 6.2 Infection cycle of the f1 phages: 1: Infection at the F pilus of the bacteria. 2: Formation of the double-stranded DNA from the single strand. 3: DNA amplification with aid of the machinery of the bacterium. 4: Stabilization of the phage DNA through phage protein pV. 5: Synthesis of phage proteins by the bacterium. 6: Assembly of the phages at the bacterial membrane and release.

molecular biology: several thousand nucleotides are able to be introduced into the bacterium via the phage genome.

The further development led to so-called phagemids, i.e. plasmids that carry the origin of replication and the assembly signal of the phage, but otherwise behave like plasmids (specific gene expression in bacteria after a start signal). With the aid of a so-called helper phage, the phagemid DNA can be replicated through the replication origin after the transfection of a bacterium, packed into particles similar to phages and secreted (Figures 6.2 and 6.4). For this the phagemid is introduced into the bacterium by means of electrotransformation; the helper phage infects the bacterium in the process. If the DNA of the phagemid carries, for example, the modified sequence of pIII, all of the phage proteins will be formed in the bacterium with the aid of the DNA of the helper phage; the replication of the helper-phage DNA can be suppressed. As a result, new phages are formed that contain phagemid DNA as DNA and a protein set that is made of mutated pIII and wild-type pIII, as well as all other proteins of the wild type. Phages of this type can infect bacteria again, but always require helper phages for proliferation [Collins, 1997].

6.1.2 COMBINATORIAL POSSIBILITIES

The first peptide mixtures in phages were described roughly simultaneously with the development of chemical libraries. The principle is quite simple: The DNA of the phage that encodes for pIII (less frequently that of pVIII) is prolongated at a suitable point by an additional piece of DNA – in the case of hexapeptides by 18 randomized nucleotides or more precisely six randomized triplet codons (referred to as NNN), i.e. instead of using a defined nucleotide sequence, every nucleotide sequence will consist of a random sequence of adenosine, thymidine, guanosine and cytosine. Any amino acid can consequently occur at any point of the hexapeptide, and all 20^6 possible hexapeptides can therefore in principle be expressed by the phages. The three triplets ATT, ATC and ACT, which are transcribed in the so-called stop codons UAA, UAG and UGA that terminate the translation, are also contained in the mixtures. Because one would like to exclude these, but nevertheless have all 20 amino acids encoded, it has been proven useful to use triplets on the DNA level that lead to reduced codons of the type NNK (N = A, C, G, or T; K = G or T), instead of NNN. Thanks to the degenerate code all of the amino acids are still encoded, but just one stop codon, which significantly reduces the frequency of chain termination (1:32 in contrast to 3:64).

Starting with hexapeptides, the path quickly led to larger fragments: from 15mer peptides through small- and medium-sized proteins up to antibodies. Large proteins are naturally not completely randomized, but instead merely contain certain randomized sequence sections, for example, for which one would like to understand the significance of individual positions that one is trying to specifically optimize by protein engineering or for which one would like to understand the significance of individual positions on the folding of the protein. The DNA of these sequence sections can now be completely or partially randomized, i.e. an $(NNK)_x$ DNA sequence replaces the corresponding sequence section represented by x amino acids.

6.1.3 THE CREATION OF A PHAGE LIBRARY AND THE AMPLIFICATION

Filamentous phages, such as the M13 phage, have been known for a long time in molecular biology, and their genome is frequently modified. The DNA that codes for the proteins pIII or pVIII can also be modified without difficulties. The resulting phages are therefore called phage type 3 or phage type 8, respectively. If the phage has, in addition to the modified gene, also that of the wild type, one speaks of phage type 33 or type 88 [Collins, 1997]. This is, for example, advantageous when large proteins are to be inserted. In this case, no longer expressing the entire pIII protein, but only its *C*-terminus, has proven itself. Phages of this type then carry intact pIII protein, which they require for infection, and a pIII protein segment that is just anchored in the phage capsid, but can no longer contribute to renewed infection. The use of phagemids has likewise proven itself. A common plasmid that carries the DNA of the modified pIII protein, a resistance gene for selection and a 'phage recognition sequence' is sufficient. A plasmid of this type can be introduced into *E. coli* by electrotransformation, and phages that carry the modified pIII proteins in addition to the pIII proteins of the wild type can be formed with the aid of a so-called helper phage. Phages of this type are then called phages type 3 + 3 (Figures 6.3 and 6.4).

How does one get to the desired diversity in the end, however? In the case of a peptide library, a cut can be made via restriction enzymes at the *N*-terminus of the pIII protein on the DNA level and a segment can be inserted (so-called ligation). This piece can be the DNA of a known protein or can be specifically produced for this purpose. The variable portion is normally introduced through randomized, synthetic primers. If one would like to integrate a whole protein on the DNA level into the phage, only one segment is usually randomized. The significance of the amino acids in the second loop of protein X for the interaction with protein Y could, for example, be clarified as follows: Protein X is expressed at the *N*-terminus of pIII, but a variable sequence is introduced instead of the natural second loop. This sequence can leave certain positions that are already known fixed, for example, or can contain all of the amino acids in a random distribution at all positions.

A further method for getting to mixtures is applied many times in the case of phage antibody libraries. The messenger (m)RNA of a patient's antibodies is obtained from the B-lymphocytes for this. It can be reverse transcribed into

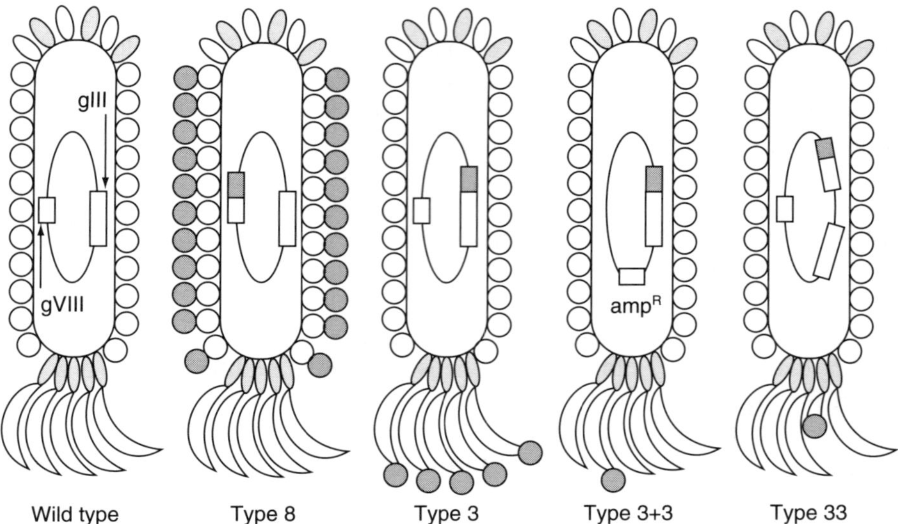

Figure 6.3 Various phages that are suitable for combinatorial methods. Depending on the type, the pVIII protein or the pIII protein is used for the mutations.

DNA by a so-called reverse transcriptase. This mixture can now be carefully cut into shape with molecular biological techniques and – packed in phagemids – expressed in phages. Methods of this type are the first tools for individual therapeutic approaches, as will be necessary in gene therapy.

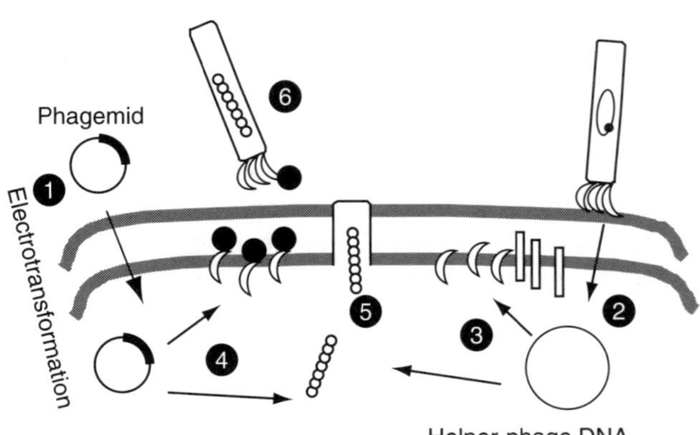

Figure 6.4 Infection cycle when work is undertaken with phagemid technology instead of phages. 1: Introduction of the phagemid through electrotransformation. 2: Infection of the bacterium with helper phages. 3: Phage envelope protein production by the bacterium (the information comes from helper phages, the information of the modified protein comes from the phagemid). 4: Amplification of the phagemid DNA by the bacterium. 5: Assembly of the phages at the bacterial membrane. 6: Release.

6.1.4 SCREENING OF THE DESIRED PHAGES

What is the special feature about this phage system that has proven itself to this degree? Phages have a few special properties. For one thing, they can reproduce themselves, i.e. pass along their properties to the next generation. Furthermore, they can be treated like solutions and thus interact with bound molecules. This characteristic is used in the search for 'desired' phages. Combinatorial methods are frequently used to identify a sequence that interacts with a certain protein, for example the epitope of a monoclonal antibody. In this case, one can coat a suitable carrier with the corresponding monoclonal antibodies and incubate it with a phage library. The phages that show a specific interaction with the antibody through their modified pIII protein segment will bind, the others will be washed away. The binding phages can likewise be detached from the carrier through more extreme washing conditions (pH value, addition of salt). After proliferation in a directed way through the infection of bacteria, they can be tested again for optimal binding (Figure 6.5). This cycle (incubation with the carrier-bound partner–washing–detachment–infection) can be repeatedly carried out; a 100- to 1000-fold enrichment of the binding phages is normally achieved per cycle, hence usually 3–4 cycles are sufficient to obtain high-affinity phages. To determine which mutants show the best interaction, the phages that are obtained in this way are proliferated; their DNA is isolated and subsequently sequenced. Because several positive clones are usually obtained, several mutants have to be separately sequenced, or so-called pool sequencing is carried out. The mutants are jointly sequenced here, and variable and constant regions are obtained as a result, i.e. DNA sections that can be varied and those that are identical in all of the binding phages and should consequently be important for the binding [Lowman, 1997].

Randomizing is frequently done again after identification of the first high-affinity sequences. One speaks of so-called libraries of the second generation in connection with this. Because one can never be certain that all possible phages have actually been formed in the first generation, there is consequently the possibility that one can identify other even better sequences. Either specific positions can be fixed here and others varied in a directed way, or else the entire sequence can be varied. This can take place, for example, through directed mutagenesis, through directed error-prone use of PCR methods for DNA amplification or through DNA shuffling (digestion and random assembly of the DNA segments) (also see Section 6.3). After renewed selection cycles, the phages with the highest affinity are again proliferated, and mutants with higher affinity are possibly obtained in the second generation that lead to molecules that bind even stronger.

Other screening methods also exist. For example, the production of antibodies or antibody segments with the aid of phages has very much proven itself. In this case, frequently only a portion of the antibody sequence is randomized, usually the variable region. A different antibody is consequently presented

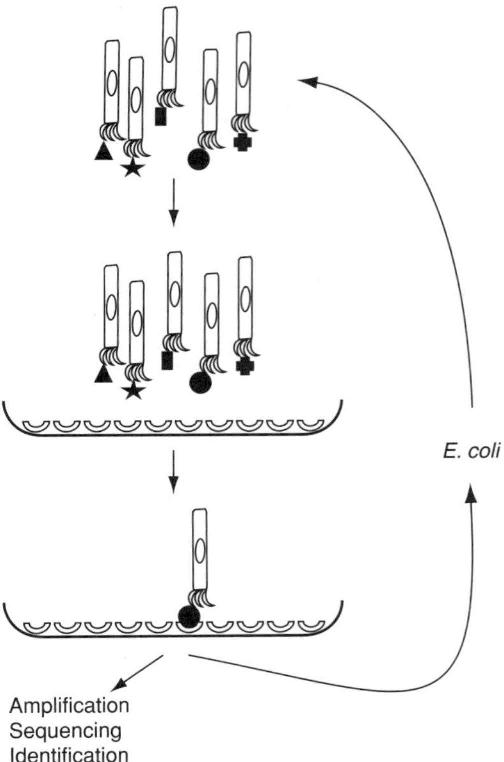

Figure 6.5 Selection cycle within the framework of the phage display technique. The phages carry different proteins: Surfaces that were coated with a binding partner are incubated with the phages; after a certain amount of time, the unbound phages are washed away; the bound ones are detached under more severe conditions, amplified in the bacterium and selected again. After 3–6 cycles, one usually obtains phages that bind well, which can then be isolated. The identification of the protein sequence takes place on the DNA level via sequencing.

in vitro on every phage, so that screening can be done with a peptide sequence, with whole cells or with the protein against which the antibody is to be aimed. The advantage consists in having the DNA that is being searched for being immediately available after identification and in being able to produce monoclonal antibodies or antibody fragments in a directed way. Naturally, phage antibodies of this type can also be used for competition with known antibodies and can, in principle, also be used for very diverse types of enzyme-linked immunosorbent assay (ELISA). If enzymes are expressed on the surfaces of the phages, screening can be done with their substrates; the following applies to this process in a very general way: all of the molecules that can be expressed on the surfaces of the phages can be screened with the respective interaction partner.

6.2 COMBINATORIAL MUTAGENESIS

In site-directed mutagenesis experiments, an amino acid of a protein sequence is normally replaced specifically by a different one. This approach can just as well be expanded for combinatorial mutagenesis by simultaneously replacing various amino acids at several positions or by replacing an amino acid at one position by several others in parallel. Furthermore, the combination of both approaches is, of course, also possible.

As is always the case when molecular biological techniques are applied, the variation of the amino acids also indirectly takes place here on the DNA level. In addition, a very effective selection method is required as we are dealing with a combinatorial approach. A direct investigation of the behavior of the mutants is possible for a specific 'individual mutation', but the pool of mutants has to be selected or sorted on some level in the case of combinatorial mutagenesis in order to obtain the desired information. This can frequently be done through the activity of the mutated protein, for example of an enzyme, which leads to modified growth behavior in the bacterium.

The general method, the principle of which has been and will be varied in diverse ways, is illustrated next with the aid of several examples.

6.2.1 SCREENING FOR ACTIVITY

The approach that is used most often is based on a screening system that selectively distinguishes active proteins from inactive proteins. Systems of this type are also used in classical mutagenesis studies, in order to distinguish transfections that are not successful (*E. coli* without protein) from successful ones (*E. coli* that express the desired protein).

As an example, proteins that play a role in the biosynthesis of amino acids can be easily studied and identified in *E. coli*, which are not able to produce this protein or can only produce it as mutants, by removing these amino acids from the culture medium. All of the *E. coli* colonies that can nevertheless grow in this culture medium have to express active variants of the biosynthesis protein. Hilvert and his group using this method investigated the properties of the enzyme chorismatemutase, a protein for the biosynthesis of aromatic amino acids [Kast *et al.*, 1996; Macbeath *et al.*, 1998] and identified the significance of various amino acids and their exchange possibilities through random mutagenesis.

Protein folding can likewise be investigated with the aid of combinatorial mutagenesis [Sauer, 1996]. Amino acids that are potentially relevant for structure and therefore relevant for function are randomly exchanged by all of the other amino acids, the 'active' proteins are subsequently identified and their sequences are determined. Depending on which amino acids are still accepted at individual positions, the actual significance of individual positions can be determined and effects based on the secondary structure can be distinguished from effects based on the tertiary structure. How one can proceed is demonstrated with structural analysis of the α-helixes of the λ repressor: Twelve positions were randomly exchanged with alanine (helix-stabilizing amino acid) or valine (helix-destabilizing amino acid) in helix 1 of the protein, which led in the case of the valine library to $2^{12} = 4096$ different proteins and in the case of the alanine library to $2^{10} = 1024$ different proteins (two of the 12 exchanged positions already represent an alanine in the natural sequence). Because active repressor proteins give bacterium immunity against an infection with a certain phage, these proteins can simply be found through the selection of bacteria that cannot be infected, which was the case with 4 % of the proteins of the valine library and 8 % of the proteins of the alanine library. Sequencing and identification of the proteins subsequently permitted more precise conclusions as to the significance of the individual positions for the structure [Gregoret and Sauer, 1998].

A simple color test is obviously an elegant possibility for distinguishing expressed enzymes that are active from inactive ones, however, this is only occasionally possible. A test system that is based on the property of RNase-T1 expressing *E. coli* to induce red colorations on suitable test plates was able to be used recently to investigate and to vary the substrate specificity of this RNA-cleaving enzyme with the aid of the random mutagenesis of six amino acids in the substrate-binding loop [Hubner *et al.*, 1999]. A mutant that exhibited an altered preference toward purine nucleotides could be identified by this approach.

6.2.2 SCREENING FOR PHYSICAL PROPERTIES

The 'green fluorescent protein' (GFP), which was originally isolated from jellyfish, has recently established itself as an important tool in molecular biology because its fluorophore is spontaneously formed from the tripeptide Ser-Tyr-Gly of the primary sequence and the protein can be introduced as a molecular biological probe into living cells. This allows the investigation of these cells by fluorometry or fluorescence microscopy (for example to study the distribution, localization and stability of a protein within a cell via fusion of the protein with GFP).

The use of GFP was limited to pH-neutral systems until recently, but the successful development of pH-dependent variants of GFP has come about with the aid of combinatorial mutagenesis. Wild-type protein was used here as a starting point, and five segments with a length of 4–5 amino acids each were randomized that potentially could influence the Ser-Tyr-Gly fluorophore according to the structure (so-called structure-directed combinatorial mutagenesis). In addition the residues were intentionally inserted, because their pH-dependent sensor properties are known. A total of 19 000 bacteria colonies were analyzed with regard to their pH dependence on fluorescence behavior and two variants were found that distinguish themselves from the wild-type sequence by nine or six amino acid exchanges, respectively. Mutants of this type are extremely interesting, because the individual cell compartments distinguish themselves by pH differences and whether proteins move, and where they move to, can now be easily studied [Miesenböck *et al.*, 1998].

6.2.3 CIRCULAR PERMUTATIONS

There is another approach for investigating structural stability, so-called circular permutation, for proteins that have spatially neighboring *N*- and *C*-termini. Theoretically, the proteins are cyclized and subsequently opened again by cleavage of peptide bonds. Experimentally, the construction of these circularly permutated proteins took place not on the protein level, but through the expression of different DNAs [Luger *et al.*, 1989]. Either active or inactive protein variants were obtained depending on the cleavage site. Cleavages that lead to inactive proteins show that sites that are sensitive to the structure are involved: protein areas that lead to active proteins even after cleavage are either not as significant or are fixed through the tertiary structure [Goldenberg and Creighton, 1983].

The technique of circular permutation has recently been successfully combined with combinatorial methods. The *N*-terminus and *C*-terminus of the desired protein, for example, are linked and cyclized here on the DNA level, and this cyclic DNA is cut in a random manner [Graf and Schachmann, 1996; Hennecke *et al.*, 1999]. This linearized DNA is cloned in the usual fashion and expressed in a suitable system, for example in *E. coli* (Figure 6.6); efficient screening is also necessary in this case in order to distinguish active proteins from inactive ones. The technique was recently used to characterize the folding

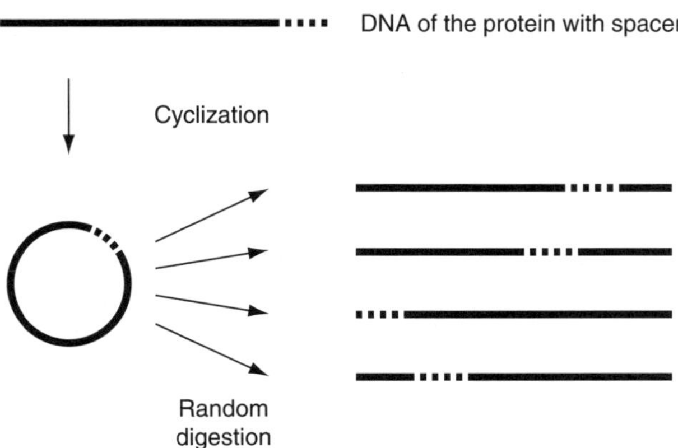

Figure 6.6 Principle of circular permutation. The DNA sequence of the desired protein is provided with a spacer and cyclized. Random digestion leads to a series of DNA pieces that are distinguished by the cutting site. The expression then leads to circularly permutated proteins.

of disulfide oxidoreductase DsbA from *E. coli* [Hennecke *et al.*, 1999]. 51 different, active mutants were able to be identified here with the aid of a β-galactosidase assay. Surprisingly, the newly generated ends were distributed over nearly the entire protein (70 % of the secondary structure elements).

6.3 *IN VITRO* SYSTEMS FOR PROTEIN BIOSYNTHESIS AND DIRECTED EVOLUTION

The protein expression takes place *in vivo*, i.e. generally in bacteria (*E. coli*) both in the case of phage display techniques as well as in the case of combinatorial mutagenesis. Limits are set on the combinatorial possibilities because of this. For example, the size of the library is limited by the introduction of the phage DNA, which only permits $10^7 - 10^9$ different components [Hoffmüller and Schneider-Mergener, 1998]. In addition, selection processes within the biosynthesis machinery of the bacterium can suppress certain mutants, not accept them as a substrate or not express them. With *in vivo* methods, therefore, selection can only be done in a positive sense, never in a negative one! This means that mutants that are 'active' meet the required criteria. On the other hand, however, no precise statements can be made as to the causes for certain mutants not being able to be found. In the simplest case, this could mean that they are not active. Although, it can just as well mean that they were degraded, that they were not expressed, that they were not able to be correctly folded or transported, or simply that they were not in existence at the generation of the diversity. How can one get around problems of this type?

For one thing, the combinatorial methods of the second generation (see the following section) allow the randomization of the molecules that were successfully identified in the first phase. Because of this it becomes possible that mutants that were not formed or not formed frequently enough could be selected from anew. The progressing alternating process of generation of diversity and selection is also called *in vitro* evolution or directed evolution [Giver and Arnold, 1998; Tobin *et al.*, 2000].

For another thing, the problems of *in vivo* systems can be tackled by systems of *in vitro* protein biosynthesis. This means that all of the elements necessary for this (mRNA, ribosome, transfer (t)RNAs) are isolated and more or less used in the test tube. The limiting problems described no longer play a role now with approaches of this type [Hoffmüller and Schneider-Mergener, 1998; Hanes and Plückthun, 1999].

6.3.1 TOOLS OF DIRECTED EVOLUTION

The libraries of the first generation are created through the use of molecular biological methods on the DNA level, i.e. primers, small DNA segments that are synthetically produced, are used as mixtures and amplified by polymerase chain reaction (PCR). Depending on the system, a translation is done, and a search is done for the active components. If the identified DNA sequence is to be further optimized, the tools for creating the second generation of libraries are made use of, also called tools of directed evolution, in which 'error-prone' PCR and so-called 'DNA shuffling' have gained the greatest degree of importance (Figure 6.7).

Error-prone PCR is understood to be the use of PCR techniques under 'unfavorable' conditions: If the optimal working conditions (Mg^{2+} concentration, temperature, enzyme concentration) of the polymerase are not maintained, the enzyme works in an incorrect fashion, does not completely copy the templates and supplies a whole set of different DNA molecules instead of identical copies that are usually desired with PCR. True to the motto 'maybe something better has come about by chance', they can then be introduced into the process again and screened. As an alternative to this, certain positions can obviously also be randomized again in a directed way if enough information is available about the system. If the process (diversification–screening–diversification–screening) is continued through several rounds, it can be compared to asexual evolution: mutants arise spontaneously; the system decides whether they are usable or not.

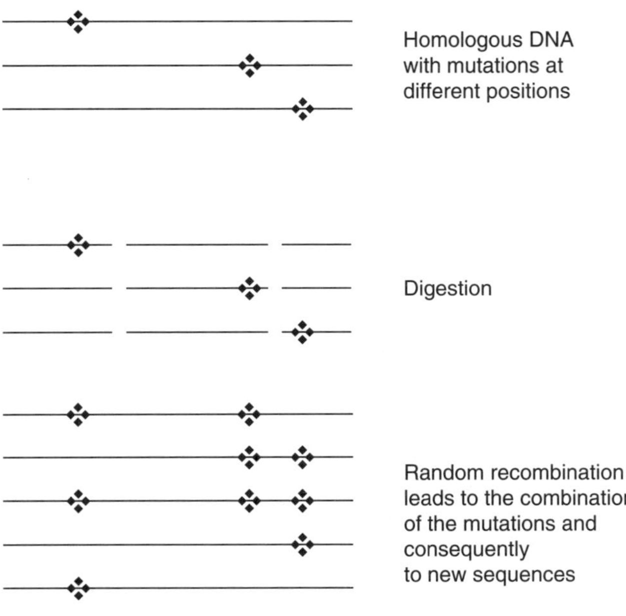

Figure 6.7 Principle of 'DNA shuffling': Homologous recombination of a pool of related sequences. The sequences that carry different mutations are cut via enzymes and amplified without the addition of a primer by PCR. Each sequence can consequently be a starting sequence for another one, which has the result of the recombination of all of the mutations.

According to this definition, the various techniques of DNA shuffling can be compared with sexual evolution, which allows exchange of DNA and rearrangement [Harris and Craik, 1998], because they run in the same way [Arnold, 1998]. Different homologous segments of the 'starting DNA' are cut with DNase and put together in a new arrangement. Mutations that were originally located on a homologue can consequently be combined with mutations of another homologue.

Meanwhile, there has been a whole series of successful studies using DNA shuffling such as improvement of the solubility of GFP, modification of the substrate specificity of tRNA synthetase and optimization of an antibody, for instance [Kuchner and Arnold, 1997; Giver and Arnold, 1998].

It was demonstrated recently that the various techniques are not to be viewed as being in competition, but can instead complement one another in an outstanding way: Defective mutants of lysozyme were optimized with both random mutagenesis and DNA shuffling. The best mutant, however, was not obtained with either method, but instead through a combination of both techniques [Jucovic and Poteete, 1998].

6.3.2 RIBOSOME DISPLAY – AN EFFECTIVE SYSTEM OF *IN VITRO* PROTEIN BIOSYNTHESIS

For *in vitro* protein biosynthesis, all of the components have to be available on an extracellular basis, they should function exactly as in the cell, and specific differences have to nevertheless exist. An mRNA copy of the DNA is created in the cell by transcription and translated at the ribosome into a protein. The mRNA–ribosome–protein complex decomposes after successful translation, and the protein is expressed according to the target location. This leads to a crucial problem *in vitro*, however, in connection with combinatorial methods: The genotype (DNA) and phenotype (protein) can no longer be assigned to each other because of the decomposition of the mRNA–ribosome–protein complex. Decisive progress was achieved, however, with the development of a system that leaves this complex intact and nevertheless forms active protein [Hanes and Plückthun, 1997]. The mRNA is used without a stop codon for this. Instead, a hairpin loop of RNA that likewise ends the translation but leaves the complex stable, in contrast to the stop codon, follows after a spacer sequence. The addition of nuclease inhibitors increases the stability of the mRNA; protein disulfide isomerases and a spacer promote folding of the protein sequence into the active conformation. The entire mRNA–ribosome–protein complex is subsequently screened for activity of the protein component, for example by polymer-bound substrates or ligands. The mRNA–ribosome–protein complexes that were bound are denatured, the mRNA is isolated, reverse transcribed into DNA by reverse transcriptase, amplified by PCR and directly sequenced or fed into the ribosome selection cycle again (Figure 6.8) and only

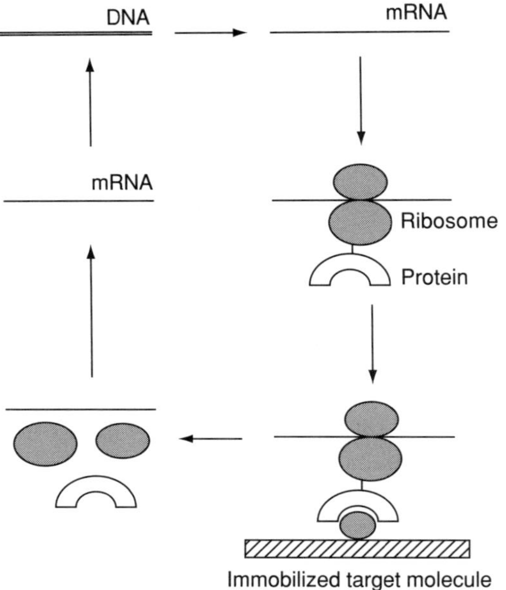

Figure 6.8 Principle of the ribosome display technique. The DNA is read as RNA (transcription) and translated *in vitro* on the ribosome into protein. This complex does not disintegrate, because the RNA was correspondingly modified and can be selected on an immobilized target molecule. The complex found in this way is subsequently destroyed, the mRNA is isolated and reverse transcribed into DNA again by reverse transcriptase.

sequenced after several rounds [Jermutus *et al.*, 1998]. The yields of intact protein are still low and are around 0.2 %, which is a great deal in comparison to earlier approaches without the modifications mentioned above where the yields were only at around 0.001 % [Hoffmüller and Schneider-Mergener, 1998]. The protein biosynthesis *in vitro* in combination with the tools of directed evolution can consequently be ranked among the methods with greatest potential for diversity (for a review see Schaffitzel *et al.* [1999]).

6.4 COMBINATORIAL METHODS BASED ON NUCLEIC ACID

The methods that have been previously described in this chapter have in common the fact that testing, screening and selection takes place on the level of the proteins, whereas the creation of diversity takes place on the level of the nucleic acids. Libraries will be explained in this section in which RNA molecules (or in a few cases also DNA molecules) are at the center of interest. It was already shown in 1990 that RNA molecules with special properties can be identified out of mixtures with the aid of suitable methods [Ellington and Szostak, 1990; Tuerk and Gold, 1990]. Since then, these techniques have been varied a number of times, optimized and verified with numerous examples. The principle, selection process and applications of RNA molecules with binding properties (aptamers) and catalytic activity (ribozymes) are described next.

6.4.1 SELEX – A STRATEGY FOR IDENTIFYING ACTIVE RNA MOLECULES

SELEX is the abbreviation for Systematic Evolution of Ligands by EXponential enrichment (Figure 6.9). The starting point here is an RNA library. RNA molecules that consist of 20–200 bases create a diversity of $4^{20} - 4^{200}$ if only the four natural nucleotides are used and thereby exceed the diversity of any chemical library. If other unusual nucleotides are used, the number of building blocks of the starting library can be expanded and – what is even more important – the possibilities for interactions can be increased. The selection takes place in a second step. What is being searched for is obviously crucial here. If one would like to obtain RNA, for example, that can bind a ligand, it can be immobilized. The RNA molecules that bind to the immobilized ligand can be fished out. The bound RNA, for example, is separated from the nonbinding RNA through filtration and then reverse transcribed into DNA by reverse transcriptase. The DNA can subsequently be amplified by PCR. This method is so potent that even individual molecules can be amplified strongly enough that they can be detected without further ado. The following cycle is run through, in principle, for the amplification of DNA by PCR:

(1) denaturation (increasing the temperature to 94 °C so that the double strands separate into individual strands);

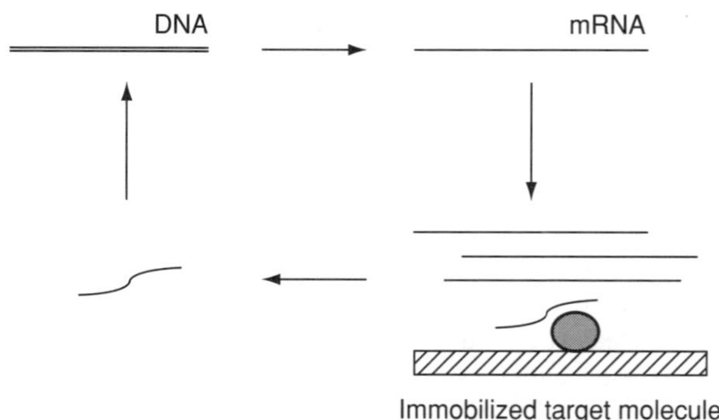

Figure 6.9 Principle of the SELEX process for searching for the desired RNA molecules. The DNA is read as RNA (transcription), selected on immobilized target molecules, isolated and reverse transcribed into DNA again by reverse transcriptase.

(2) renaturation/annealing (lowering the temperature to 55 °C and addition of so-called primers (complementary, short pieces of DNA));
(3) extension (increasing the temperature to 72 °C and extension of the primer DNA by a thermostable polymerase, because of which double-stranded DNA is formed).

The DNA that is amplified in this way is reverse transcribed into RNA again by T7 RNA polymerase and can be screened again. One usually requires three to eight cycles of this type in order to obtain strongly binding and selective RNA molecules. The cycle run is schematically depicted in Figure 6.9. The identification of the RNA molecules is possible without any problem, because they can be sequenced in each case on the DNA level. It then becomes clear whether identical, homologous RNA molecules have been selected or whether different RNA sequences were able to bind to the target molecule. This distinction can be compared with monoclonal and polyclonal antibodies. The molecular mechanism of the RNA–ligand interaction can then be elucidated by structure determination methods, such as nuclear magnetic resonance (NMR) or X-ray structure analysis, for instance. Various examples have already been published with regard to this (Figure 6.10).

6.4.2 APTAMERS – RNA WITH RECOGNITION PROPERTIES

RNA molecules that can recognize other molecules are called aptamers. Aptamers that recognize small molecules, pharmaceutical drugs, amino acids and

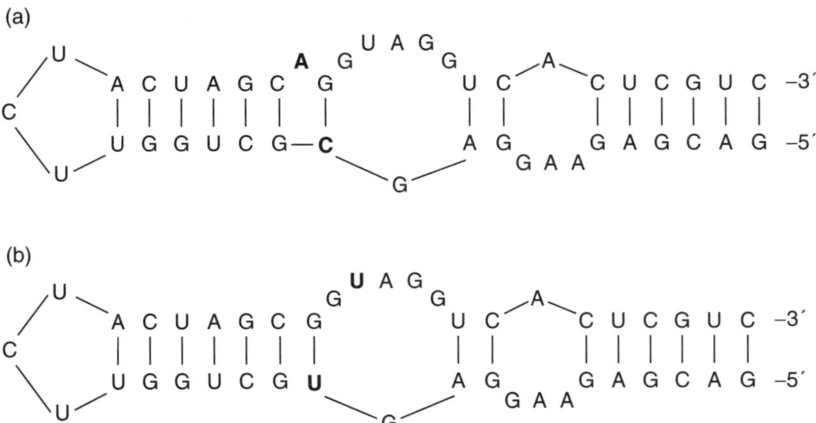

Figure 6.10 RNA aptamers that specifically bind the amino acids arginine (a) and citrulline (b). Both aptamers consist of 44 nucleotides. The various positions are marked in boldface [Burgstaller *et al.*, 1995].

lipids, aptamers that bind peptides and proteins and also aptamers that recognize DNA have been described up to now [Famulok and Jenne, 1998; Gold, 1999; Brody and Gold, 2000; Famulok *et al.*, 2000; Hermann and Patel, 2000; Kusser, 2000]. Aptamers can be compared with antibodies in some respects, at least with regard to their area of use: Aptamers are suitable in the area of diagnosis, as tools in biochemistry and possibly also as therapeutics in the future. They frequently consist of 20–40 building blocks, which is why the molecules have a molecular weight of 5–10 kDa. They are consequently considerably smaller than antibodies (immunoglobulin G (IgG) = 150 kDa) and their fragments [Osborne *et al.*, 1997]. Furthermore, they are protease resistant, and exist in the body in diverse ways. Nuclease resistance can be achieved through the use of unusual nucleotide building blocks which is possible in a relatively simple way, in contrast to the proteins (see Section 6.4.4). Other advantages of aptamers are their low immunogenicity and much lower manufacturing costs in comparison to the production of antibodies [Heus, 1997].

Speculation exists as to other possibilities for the use of aptamers in the future. Suitable DNA could be introduced into certain cells by gene therapy methods. The cells themselves would then produce the RNA aptamers, so they could be directly at the site of action and the difficult transportation there (drug delivery) could consequently be eliminated.

It is to be noted that aptamers were also developed based on DNA. The examples of this are limited because DNA, in contrast to RNA, seldom exists in the form of a single strand and hardly develops structures that can interact with ligands [Breaker, 1997].

6.4.3 RIBOZYMES – RNA WITH CATALYTIC ACTIVITY

Besides research on RNA molecules with binding properties, another research area concentrates on the search for RNA molecules with catalytic properties, so-called ribozymes. Presently seven natural ribozymes have been found which can be divided into two groups: so-called large ribozymes, such as Group I and Group II introns, and the bacterial RNAse P, the latter of which creates 5′-phosphate and 3′-hydroxyl ends during the transesterification or hydrolysis of phosphodiester bonds; and the small ribozymes that contain 40–80 nucleotides and to which the hammerhead, hairpin, delta and VS ribozymes belong [Pan, 1997]. 2′,3′-Cyclic phosphate and a 5′-hydoxyl group are formed in the reactions of the phosphodiester bond catalyzed by them.

The first combinatorial work on ribozymes took the known ribozymes as a starting point, and an attempt was made through variation of individual or larger areas to change their catalytic properties: variation of the selectivity, the ion sensitivity and improvement of the catalytic properties [Kuchner and Arnold, 1997]. More recent work shows that completely new ribozymes can be created and identified via RNA libraries that allow one to significantly expand the catalytic properties of the ribozymes. Ribozymes have been identified which act as RNA ligases, as RNA polynucleotide kinases, and as aminoacyl transferases which consequently imitate the reaction at the ribosome. Even ribozymes that catalyze a Diels–Alder reaction have been created (see review by Famulok and Jenne [1998]).

How does one go about the search for ribozymes? In principle, the procedure runs in an analogous fashion to the SELEX process described in Section 6.4.1. A distinction can then be made between indirect and direct methods for the separation of RNA molecules with and without catalytic properties. In the indirect procedure, one proceeds as in the search for catalytic antibodies: A molecule is synthesized that represents the transition state of the reaction to be catalyzed. This molecule is then used for screening, because one assumes that the catalytic capabilities of the antibodies or ribozymes are based on stabilizing the transition state of a reaction through binding. In the direct methods, the search is undertaken with substrates and RNA molecules that can bind the substrates are separated by physical methods from RNA molecules that cannot transform the substrates. This can, for example, take place with the aid of electrophoresis for which RNA–substrate complexes have other mobilities than pure RNA. As an alternative, the substrate can be provided with a label (e.g. biotin), which likewise allows selection [Pan, 1997].

6.4.4 MODIFIED RNA LIBRARIES

An attempt is being made to modify RNA molecules to increase nuclease stability (enzymes that degrade oligonucleotides), to increase interaction possibilities and to improve use in diagnostics and therapy. In principle, two possibilities are described for this: direct use of modified bases and modification by chemical reactions after selection [Eaton, 1997].

The former possibility has the advantage that the modification is already taken into consideration in the selection, but also the disadvantage that only those modifications that are accepted by the T7 RNA polymerase are possible. It has been shown here that position 5 of the pyrimidines, position 8 of the purines and position 2 of all nucleotides can be varied the easiest. A selection of synthetic nucleotides that can still be used by T7 RNA polymerase is summarized in Figure 6.11. Bases of this type additionally allow hydrophobic, polar or ionic interactions and thereby increase the potential of the RNA binding.

In the second method, which is also called 'post-SELEX' modification, selection and amplification are undertaken with natural nucleotides. The RNA mixture obtained in this way is produced in fairly large quantities and then chemically modified. After that, a test has to be done again as to which of the modified RNA molecules still have the desired properties, because the modification can have both positive and negative influences. The subsequent

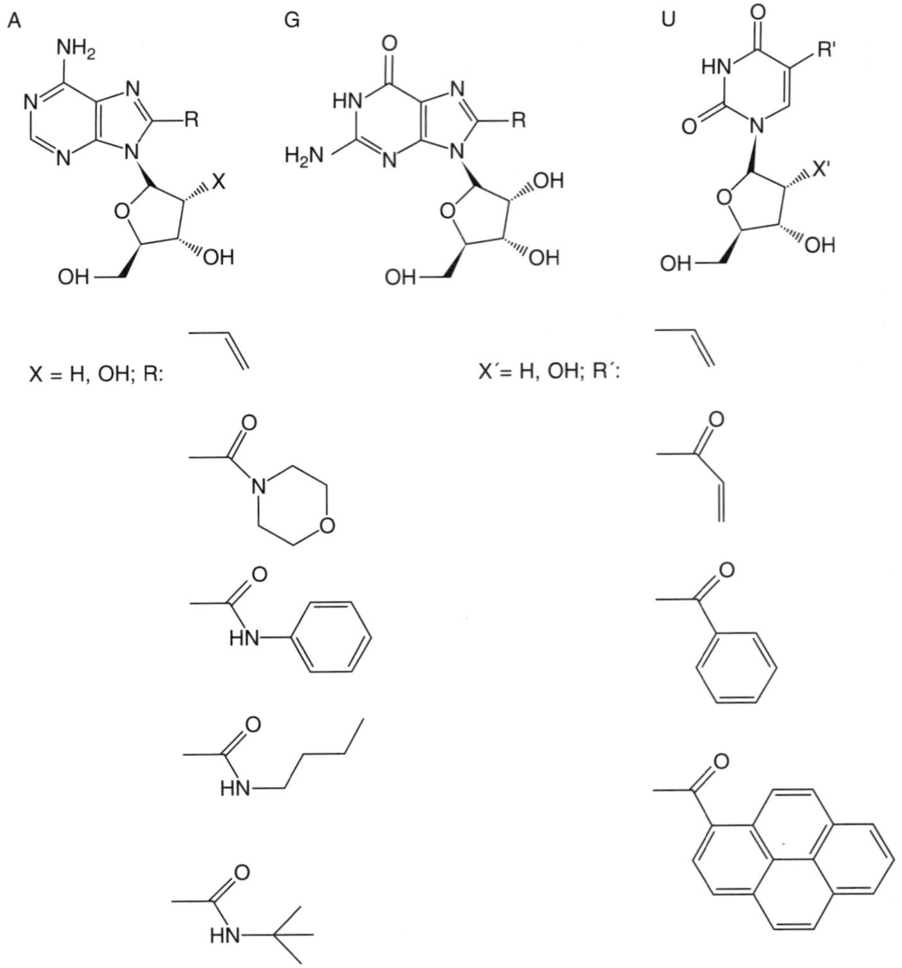

Figure 6.11 Examples of nonnatural analogues of the nucleotides 2′-deoxyuridine, uridine, 2′-deoxyadenine, adenine and guanosine that were modified at position 5 or position 8. These unusual bases are still recognized by the T7 DNA polymerase and can therefore be built into aptamers [Eaton, 1997].

identification of the most active RNA molecules is a challenge with this method; some success was achieved here by electrospray ionization (ESI) mass spectrometry (see Section 7.4).

REFERENCES

Arnold, F. H. (1998). Design by directed evolution. *Acc. Chem. Res.* **31**, 125–131.
Breaker, R. R. (1997). DNA aptamers and DNA enzymes. *Curr. Opin. Chem. Biol.* **1**, 26–31.

Brody, E. N. and Gold, L. (2000). Aptamers as therapeutic and diagnostic agents. *Rev. Mol. Biotechnol.* **74**, 5–13.

Burgstaller, P. Kochoyan, M. and Famulok, M. (1995). Structural probing and damage selection of citrulline- and arginine-specific RNA aptamers identify base positions required for binding. *Nucleic Acids Res.* **23**, 4769–4776.

Collins, J. (1997). Phage display. *Ann. Rev. Comb. Chem. Mol. Div.* **1**, 210–262.

Cwirla, S. E. Peters, E. A. Barrett, R. W. and Dower, W. J. (1990). Peptides on phage: a vast library of peptides for identifying ligands. *Proc. Natl. Acad. Sci. USA*, **87**, 6378–6382.

Eaton, B. E. (1997). The joys of *in vitro* selection: chemically dressing oligonucleotides to satiate protein targets. *Curr. Opin. Chem. Biol.* **1**, 10–16.

Ellington, A. D. and Szostak, J. W. (1990). *In vitro* selection of RNA molecules that bind specific ligands. *Nature*, **346**, 818–822.

Famulok, M. and Jenne, A. (1998). Oligonucleotide libraries – variatio delectat. *Curr. Opin. Chem. Biol.* **2**, 320–327.

Famulok, M. Mayer, G. and Blind, M. (2000). Nucleic acid aptamers – from selection *in vitro* to applications *in vivo*. *Acc. Chem. Res.* **33**, 591–599.

Giver, L. and Arnold, F. H. (1998). Combinatorial protein design by *in vitro* recombination. *Curr. Opin. Chem. Biol.* **2**, 335–338.

Gold, L. (1999). Globular oligonucleotide screening via the SELEX process: aptamers as high-affinity, high-specificity compounds for drug development and proteomic diagnostics. *Comb. Chem. Technol.* 389–403.

Goldenberg, D. P. and Creighton, T. E. (1983). Circular and circularly permuted forms of bovine pancreatic trypsin inhibitor. *J. Mol. Biol.* **165**, 407–413.

Graf, R. and Schachmann, H. K. (1996). Random circular permutation of genes and expressed polypeptide chains: application of the method to the catalytic chains of aspartate transcarbamoylase. *Proc. Natl. Acad. Sci. USA*, **93**, 11 591–11 596.

Gregoret, L. M. and Sauer, R. T. (1998). Tolerance of a protein helix to multiple alanine and valine substitutions. *Fold. Des.* **3**, 119–126.

Hanes, J. and Plückthun, A. (1997). *In vitro* selection and evolution of functional proteins by using ribosome display. *Proc. Natl. Acad. Sci. USA*, **94**, 4937–4942.

Hanes, J. and Plückthun, A. (1999). In vitro selection methods for screening of peptide and protein libraries. *Curr. Top. Microbiol. Immunol.* **243**, 107–122.

Harris, J. L. and Craik, C. S. (1998). Engineering enzyme specificity. *Curr. Opin. Chem. Biol.* **2**, 127–132.

Hennecke, J. Sebbel, P. and Glockshuber, R. (1999). Random circular permutation of DsbA reveals segments that are essential for protein folding and stability. *J. Mol. Biol.* **286**, 1197–1215.

Hermann, T. and Patel, D. J. (2000). Adaptive recognition by nucleic acid aptamers. *Science*, **287**, 820–825.

Heus, H. A. (1997). RNA aptamers. *Nat. Struct. Biol.* **4**, 597–600.

Hoffmüller, U. and Schneider-Mergener, J. (1998). *In vitro* evolution and selection of proteins: ribosome display for larger libraries. *Angew. Chem. Int. Ed. Engl.* **37**, 3241–3243.

Hubner, B. Haensler, M. and Hahn, U. (1999). Modification of ribonuclease T1 by random mutagenesis of the substrate binding segment. *Biochemistry*, **38**, 1371–1376.

Jermutus, L. Ryabova, L. A. and Plückthun, A. (1998). Recent advances in producing and selecting functional proteins by using cell-free translation. *Curr. Opin. Biotechnol.* **9**, 534–548.

Johnsson, K. and Ge, L. (1999). Phage display of combinatorial peptide and protein libraries and their applications in biology and chemistry. *Curr. Top. Microbiol. Immunol.* **243**, 87–105.

Jucovic, M. and Poteete, A. R. (1998). Delineation of an evolutionary salvage pathway by compensatory mutations of a defective lysozyme. *Protein Sci.* **7**, 2200–2209.

Kast, P. Asif-Ullah, M. Jiang, N. and Hilvert, D. (1996). Exploring the active site of chorismate mutase by combinatorial mutagenesis and selection: the importance of electrostatic catalysis. *Proc. Natl. Acad. Sci. USA*, **93**, 5043–5048.

Katz, B. A. (1997). Structural and mechanistic determination of affinity and specificity of ligands discovered or engineered by phage display. *Ann. Rev. Biophys. Biomol. Struct.* **26**, 27–45.

Kuchner, O. and Arnold, F. H. (1997). Directed evolution of enzyme catalysts. *Trends Biotechnol.* **15**, 523–530.

Kusser, W. (2000). Chemically modified nucleic acid aptamers for *in vitro* selections: evolving evolution. *Rev. Mol. Biotechnol.* **74**, 27–38.

Lowman, H. B. (1997). Bacteriophage display and discovery of peptides leads for drug development. *Ann. Rev. Biophys. Biomol. Struct.* **26**, 401–424.

Luger, K. Hommel, U. Herold, M. Hofsteenge, J. and Kirschner, K. (1989). Correct folding of circularly permuted variants of a beta alpha barrel enzyme *in vivo*, *Science*, **243**, 206–210.

Macbeath, G. Kast, P. and Hilvert, D. (1998). Exploring sequence constraints on an interhelical turn using *in vivo* selection for catalytic activity. *Protein Sci.* **7**, 325–335.

Miesenböck, G. De Angelis, D. A. and Rothman, J. E. (1998). Visualizing secretion and synaptic transmission with pH-sensitive green fluorescent proteins. *Nature*, **394**, 192–195.

Osborne, S. E. Matsumura, I. and Ellington, A. D. (1997). Aptamers as therapeutic and diagnostic reagents: problems and prospects. *Curr. Opin. Chem. Biol.* **1**, 5–9.

Pan, T. (1997). Novel and variant ribozymes obtained through *in vitro* selection. *Curr. Opin. Chem. Biol.* **1**, 17–25.

Sauer, R. T. (1996). Protein folding from a combinatorial perspective. *Fold. Des.* **1**, R27–R30.

Schaffitzel, C. Hanes, J. Jermutus, L. and Plückthun, A. (1999). Ribosome display: an *in vitro* method for selection and evolution of antibodies from libraries. *J. Immunol. Methods*, **231**, 119–135.

Scott, J. K. and Smith, G. P. (1990). Searching for peptide ligands with an epitope library. *Science*, **249**, 386–390.

Tobin, M. B. Gustafsson, C. and Huisman, G. W. (2000). Directed evolution: the 'rational' basis for 'irrational' design. *Curr. Opin. Struct. Biol.* **10**, 421–427.

Tuerk, C. and Gold, L. (1990). Systematic evolution of ligands by exponential enrichment: RNA ligands to bacteriophage T4 DNA polymerase. *Science*, **249**, 505–510.

Winter, J. (1996). Bacteriophage display libraries. *Protein Eng.*, 349–367.

7 The Analysis of Libraries and Arrays

The combinatorial methods put very high demands on analysis techniques such as, for instance:

- High sensitivity for analysis of the smallest quantities of compounds.
- A measurement principle that is preferably nondestructive so that no 'compound loss' is experienced because of the analysis.
- Automatability, as well as short analysis times and minimal sample preparation, in order to provide a high sample throughput.
- Automatic and/or quick data analysis, in order to rapidly obtain the desired results.

None of the analysis techniques that are currently available is in a position at the moment to perfectly meet all of the requirements, so the techniques are frequently used in parallel or in a coupled way. Individual analysis techniques can, however, fully meet the demands for specific areas, because the requirements that are described no longer exist in part, or have a different weight. The advantages and disadvantages of the various methods are explained in more detail next (also see the general review articles of Egner and Bradley [1997], as well as of Gallop and Fitch [1997]).

7.1 CHROMATOGRAPHIC METHODS

Chromatographic procedures are the most widespread analysis methods in the area of combinatorial research. High-performance liquid chromatography (HPLC) or micro-HPLC is mentioned before all the others here [Griffey *et al.*, 1998; Kirkland, 2000]. But a multitude of other methods are also used, especially capillary electrophoresis (CE) [Gaus *et al.*, 1999; Ma *et al.*, 2000] and supercritical fluid chromatography (SCF) [Coleman, 1999; Berger *et al.*, 2000; Berger and Wilson, 2000]. The chromatographic procedures alone are merely for the separation of compounds or compound mixtures, however, and only the coupling with a suitable detection system makes the actual analysis possible. The ultraviolet (UV) detection of compounds is in very widespread use in connection with this, because the purity of syntheses can be simply checked with chromatographic techniques, but the different absorption behavior of the various compound classes has to be taken into account here.

Evaporative light-scattering detection (ELSD) has recently become a powerful alternative for the detection and quantitation of combinatorial libraries of low molecular weight organic compounds, because the response of the detector is independent of the sample chromophore. This makes it well-suited to the rapid quantitation of mixtures of dissimilar solutes and to the detection of impurities that absorb poorly in the UV [Kibbey, 1996; Hsu *et al.*, 1999; Fang *et al.*, 2000].

Chemiluminescent nitrogen detection (CLND) represents another recent alternative that allows the rapid and accurate quantitation of low molecular weight organic compounds down to low picomolar levels [Taylor *et al.*, 1998; Lewis *et al.*, 2000; Shah *et al.*, 2000].

Because the products of a combinatorial synthesis or the active compounds in a biological test are usually completely unknown, obtaining as much structural information as possible about the compounds separated by chromatography is required to an increased degree. This is why an increased effort has been undertaken in the past few years to hyphenate spectrometric or spectroscopic techniques to chromatographic procedures. This applies above all to infrared spectroscopy (IR), nuclear magnetic resonance spectroscopy (NMR) and mass spectrometry (MS).

7.2 INFRARED SPECTROSCOPY

IR spectroscopy is very well suited to the investigation of polymers or polymer-bound molecules [Gremlich, 1999]. Its main areas of use are the identity check of compounds through comparison with spectra databases, as well as the detailed investigation of solid phase reactions of know compounds, because data obtained through IR spectroscopy do not permit a direct structural analysis of unknown compounds.

At the beginning of the 1970s, a dispersive instrument that was common at that time was successfully used to investigate various anchor types on polystyrene resin and their reaction with saccharide building blocks [Frechet and Schuerch, 1971]. Relatively large quantities of resin had to be ground in a destructive fashion to allow the (time consuming) measurement of polymer samples as KBr pellets, which limited the possibilities for use of IR spectroscopy in the area of combinatorial chemistry.

IR spectroscopy has gained a great deal of significance in the past few years through the introduction of the Fourier transform (FT) technique: modern FT-IR devices allow quick measurements and have high sensitivity. Polymer-bound compounds and their reactions can be investigated in real time both qualitatively and quantitatively in the picomolar range. Yan *et al.* for the first time in 1995 succeeded in following reactions on individual polymer beads [Yan *et al.*, 1995]. Their experimental setup consisted of a spectrophotometer that they coupled to an IR microscope, in order to focus the radiation on a polymer bead. The measurements that were carried out in the transmission mode had a

Figure 7.1 Different approaches to the IR spectroscopic investigation of individual beads. (a) Standard measurement of a bead in the transmission mode. (b) The resolution can be increased by flattening the beads [Yan and Kumaravel, 1996]. (c) The attenuated total reflection (ATR) measurement technique permits surface reactions to be followed (the white region schematically represents the measurement area) [Yan *et al.*, 1996a].

relatively low resolution because the beam had a long, irregular path due to the spherical shape of the polymer beads. The results could, however, be significantly improved by 'flattening' the polymer beads (Figure 7.1) [Yan and Kumaravel, 1996]. Among other things, the catalytic oxidation of alcohols to aldehydes on individual beads was able to be followed in this way in real time (Figure 7.2) [Yan *et al.*, 1996b] and the influence of the polymer on the kinetics of solid phase reactions was able to be studied (Figure 7.3) [Li and Yan, 1998].

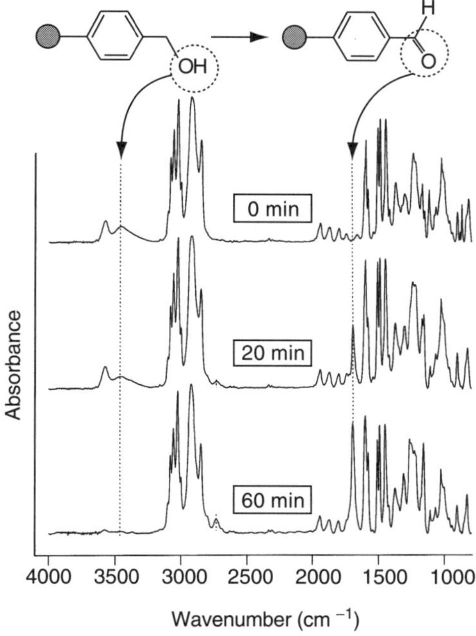

Figure 7.2 Study of the catalytic oxidation of a primary alcohol to an aldehyde on individual, flattened beads in real time by means of FT-IR spectroscopy according to Yan *et al.*, [1996b]. The signal of the hydroxyl stretch (with and without hydrogen bridges) of the alcohol in the range around 3 424 cm^{-1} disappears.In contrast to this, the signal of the carbonyl stretch of the aldehyde that is formed appears at 1695 cm^{-1}. (The spectra were kindly made available by Novartis, Basel, Switzerland.)

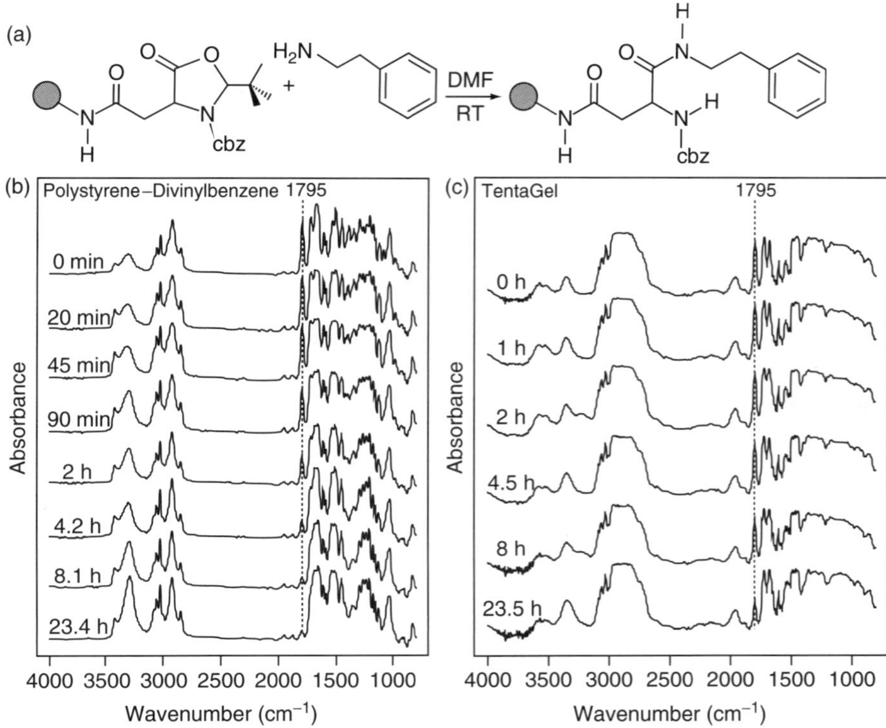

Figure 7.3 Investigation of polymer influence on the kinetics of solid phase reactions by means of FT-IR spectroscopy according to Li and Yan [1998]. (a) Ring opening reaction investigated. (b) Spectra of the time-dependent measurements on a polystyrene–divinylbenzene bead. (c) Spectra of the time-dependent measurements on a TentaGel bead. The speed of the ring opening reaction, which can be studied in light of the decrease of the IR band at $1795\,cm^{-1}$, is 18 times greater for the polystyrene–divinylbenzene polymer in the case at hand than for the TentaGel polymer. (The spectra were kindly made available by Novartis, Basel, Switzerland.)

Another measurement technique, so-called attenuated total reflection (ATR) permits surface reactions to be followed (Figure 7.4) [Yan *et al.*, 1996a; Haap *et al.*, 1998]. A special ATR lens is brought into direct physical contact with the sample for this. The IR beam penetrates into the sample ($\approx 2\,\mu m$); it is reflected with attenuated intensity and subsequently analyzed. Much smaller sample quantities ($\approx 100\,fmol$) are able to be detected with this technique than is the case in the investigation of individual beads in the transmission mode, as a comparison of the effectively investigated sample volumes illustrates (see Figure 7.1).

Approaches to the on-line coupling of chromatographic methods and FT-IR spectroscopy were also pursued based on the relatively short measurement times in the second range; the LC–FT-IR coupling is mainly of significance

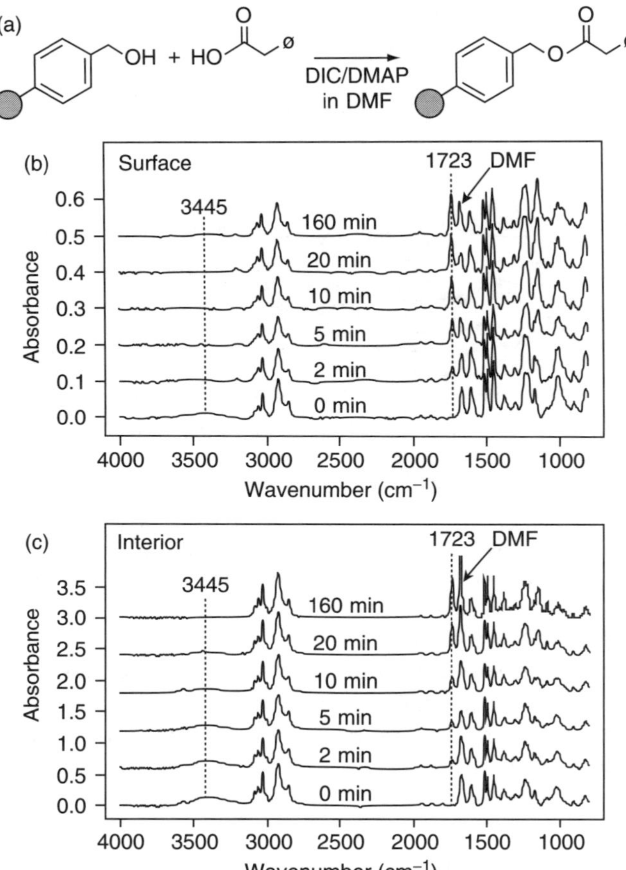

Figure 7.4 FT-IR spectroscopic investigation of an esterification by means of the ATR technique according to Yan *et al.* [1996a]. (a) Reaction being investigated. (b) Spectra of the time-dependent measurements by means of the ATR technique on the surface of a bead. (c) Spectra of the time-dependent measurements on a flattened bead in the transmission mode.The reaction at hand can be studied in light of the decrease of the hydroxyl signal at 3445 cm^{-1} and the increase of the carbonyl signal of the ester at 1723 cm^{-1}. Ultimately, 130 fmol of compound was measured with the use of the ATR technique; in contrast, in the case of the transmission measurement around 4000–fold of quantity (500 pmol) was analyzed. (The spectra were kindly made available by Novartis, Basel, Switzerland.)

for combinatorial chemistry. Because the solvent usually exerts a disturbing influence on IR measurements, direct on-line coupling by means of flow cells is only suitable for special areas of use. Various semi on-line methods, in which the samples are deposited on KCl powder (diffuse reflection), metal plates (reflection–absorption) or transparent carriers (transmission), have a much broader range of application (Figure 7.5). Very diverse techniques are used

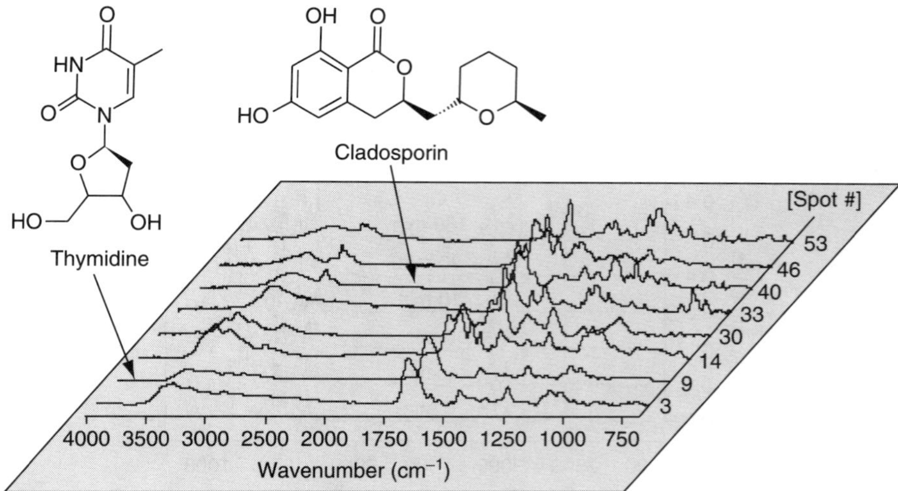

Figure 7.5 Semi on-line LC–IR coupling for the separation and detection of thymidine and cladosporin.The fractions were deposited as spots on metal plates and subsequently individually measured. (The spectra were kindly made available by Novartis, Basel, Switzerland.)

here for vaporizing the solvent; commercially available devices are chiefly based on compressed air or ultrasound at the moment.

Somsen *et al.* [1998] offer a comprehensive overview of the coupling of liquid chromatography and FT-IR.

7.3 NMR SPECTROSCOPY

A great deal more structural information can be obtained by NMR spectroscopy than is the case with most other analytical techniques. Based on the low sensitivity, relatively large, homogenous samples, a great deal of time as well as deuterated, and therefore expensive, solvents were required up to a few years ago to get good results. NMR spectroscopy was therefore only of conditional use within the scope of combinatorial chemistry, because neither direct investigation of low substance quantities on polymeric support nor automatic and quick analysis of low quantitities of a multitude of compounds in solution was possible.

Various work groups have recently succeeded in solving these problems in principle. Thus, today NMR spectroscopy has high significance in the area of combinatorial chemistry, but there is still a high potential for further developments. Review articles by Keifer offer an excellent overview of the most recent developments in this area of NMR spectroscopy [Keifer, 1998; Keifer, 1999].

7.3.1 NMR ON A POLYMERIC SUPPORT

Gel-Phase NMR

The quality of NMR data is determined by the line width (spectral resolution) of the signals. The restriction of degrees of freedom exerts a negative influence here, which is why polymers are usually not investigated in the solid state, but instead in a swollen form ('gel phase'). Furthermore, the physical homogeneity of the sample is of great importance for the line shape, which is why dissolved compounds provide sharper signals in principle than heterogeneous polymer/ solvent mixtures. The underlying physical measurement variable of the compounds is called magnetic susceptibility and is dependent on the respective nucleus. This led to gel-phase NMR spectroscopy mainly being applied to ^{13}C [Schaefer, 1971; Manatt *et al.*, 1980b] and also in part to ^{19}F [Manatt *et al.*, 1980a; Manatt *et al.*, 1980b; Shapiro *et al.*, 1996] or to ^{31}P [Johnson and Zhang, 1995], because this contribution to the line broadening turns out to be relatively low for these nuclei, in contrast to ^{1}H ($\approx$50 Hz for ^{13}C in contrast to $\approx$ 200 Hz for ^{1}H). This is also promoted by the fact that the chemical shift of ^{13}C, in contrast to ^{1}H, extends over a very broad range. The natural abundance of ^{13}C is only 1.1 %, which is why ^{13}C-enriched samples are frequently used to increase the sensitivity of the experiments or to shorten the measurement times [Look *et al.*, 1994; Barn *et al.*, 1996; Gordeev *et al.*, 1996].

Because the resolution that can be achieved by NMR spectroscopy in solution is better than would ever be possible with gel-phase NMR spectroscopy (see previous paragraph), so-called 'cleave and analyze' procedures were applied (strictly speaking, merely conventional procedures in solution after cleavage of the compounds from the polymer), on the one hand [DeWitt *et al.*, 1993; Forman and Sucholeiki, 1995] and, on the other hand, various soluble polymer systems based on polyoxyethylene, polyethylene glycol and dendrimers were developed [Leibfritz *et al.*, 1978; Han *et al.*, 1995; Han and Janda, 1996; Kim *et al.*, 1996].

Magic Angle Spinning (MAS) NMR

It was recognized in 1970 that the line broadening that can be traced back to the heterogeneity of the sample, or its magnetic susceptibility, could be avoided by magic angle spinning (MAS) [Doskocilova *et al.*, 1970]. This is understood to be the mathematically deducable spinning of the sample at an angle of 54.7° to the magnetic field B_0. The method therefore also came into use a few years later to investigate swollen polymers by means of ^{13}C NMR spectroscopy and even ^{1}H NMR spectroscopy. It distinguished itself in comparison to customary gel-phase NMR spectroscopy by a reduction in the need for the compound and improved resolution, even if the measurement times were still high

[Doskocilova *et al.*, 1974; Doskocilova *et al.*, 1978; Stoever and Frechet, 1989]. High-resolution ^{1}H NMR spectroscopy of swollen polymers with a bandwidth < 8 Hz was first achieved in 1994 by optimization of the probe – since then even bandwidths of 1 Hz have been documented [Fitch *et al.*, 1994; Kempe *et al.*, 1997]. The magnetic susceptibility of the materials that were used was adapted to the special MAS requirements and the geometry of the probe was chosen in such a way that the complete sample could be brought into the active region of the NMR coil; because of this a significant increase in resolution and sensitivity was achieved. One also speaks of nanoprobes, in connection with this. The three sample components, polymer/coupled compound/solvent, also have to be coordinated with each other to the maximum extent possible for optimal results. The solvent that is chosen should not only be able to swell the polymer well, but also to solvate the coupled compound well in order to avoid line broadening through the restriction of degrees of freedom [Keifer, 1996].

Even the analysis of 'one bead one compound' libraries with bead diameters of 100 μm and a loading of $\leq$800 pmol has been achieved with ^{1}H and ^{13}C NMR spectroscopy because of the high sensitivity of the nanoprobes in combination with the partial ^{13}C labeling of the samples [Sarkar *et al.*, 1996]. Special isotope-filtered measurement techniques have to be used, however, based on the relatively strong interfering signals due to impurities as well as solvent and polymer contributions, in order to get signals that can be analyzed in the subnanomolar range. At the moment, the informational content that can be obtained is therefore not adequate to perform *de novo* structure determination and thus the routine application is still questionable. One potential area of application, however, would be the encoding of 'one bead one compound' libraries through clearly distinguishable ^{13}C-tagged linkers or through the direct ^{13}C tagging of individual building blocks.

A multitude of very diverse one- and two-dimensional NMR experiments have also been carried out recently to investigate polymeric supports with MAS NMR (Table 7.1); this will not be looked into in more depth, however.

7.3.2 NMR IN SOLUTION

Combinatorial chemistry places a series of general conditions on potential analysis methods (see previous sections) that conventional NMR spectroscopy in solution cannot fulfill in the required manner:

(1) The samples are dissolved in protonated solvents. This generates strong background signals, which makes a useful interpretation of the spectra impossible.
(2) The compounds frequently exist as mixtures, and the sample volume is too small to be capable of being analyzed with conventional NMR spectroscopy in solution.

Table 7.1 Overview of common one and two-dimensional NMR experiments investigating polymeric supports, or polymer-bound compounds, by means of magic angle spinning according to Keifer [1998].

Detected nucleus	Experiment	Abbreviation
^{1}H	Correlation spectroscopy	COSY
^{1}H	Double quantum filtered correlation spectroscopy	DQCOSY
^{1}H	Total correlation spectroscopy	TOCSY
^{1}H	Nuclear Overhauser effect spectroscopy	NOESY
^{1}H, ^{15}N	Heteronuclear multiple quantum coherence	HMQC
^{13}C	Attached proton test	APT
^{13}C	Distortionless enhancement by polarization transfer	DEPT
^{13}C	^{13}C–^{1}H heteronuclear correlation	HETCOR

(3) The samples are usually in vessels, for example microtiter plates, that are not compatible with conventional NMR spectroscopy.
(4) The sample throughput is too high for conventional NMR analysis, because the recording and evaluation of a high-quality spectrum takes a relatively large amount of time.

Deuterated solvents – and therefore very expensive ones – or else special measurement techniques for suppressing the solvent signals have been used for quite a while to solve problem (1) (Figure 7.6).

Problem (2) was mainly tackled through the successful hyphenation of chromatographic techniques with NMR spectroscopy; above all the HPLC–NMR coupling is of great importance in the area of combinatorial chemistry (Figure 7.7(A)) [Strohschein *et al.*, 1997; Albert *et al.*, 1998]. Problem (3) can also be solved relatively simply in connection with this through the upstream connection of an autosampler (Figure 7.7(B)). A few Micrograms per peak are already sufficient for useful results on the grounds of development of very sensitive flow probes in combination with NMR devices that have become more and more powerful [Sidelmann *et al.*, 1995]. More detailed information on the HPLC–NMR coupling can be found in several review articles [Korhammer and Bernruether, 1996; Peng, 2000; Stockman, 2000].

The coupling of NMR spectroscopy with capillary electrophoresis (CE–NMR) was able to likewise be realized in the past few years [Wu *et al.*, 1994; Albert, 1995; Pusecker *et al.*, 1998]. The small diameter of the capillaries puts a great deal of demands on the design of the probes, however, but allows the analysis of compound quantities in the low nanogram range [Olson *et al.*, 1996].

HPLC–NMR and CE–NMR measurements can, in principle, be carried out both in the flow-through mode (short time in the probe, for large sample quantities), as well as in the stopped-flow mode (the time in the probe is

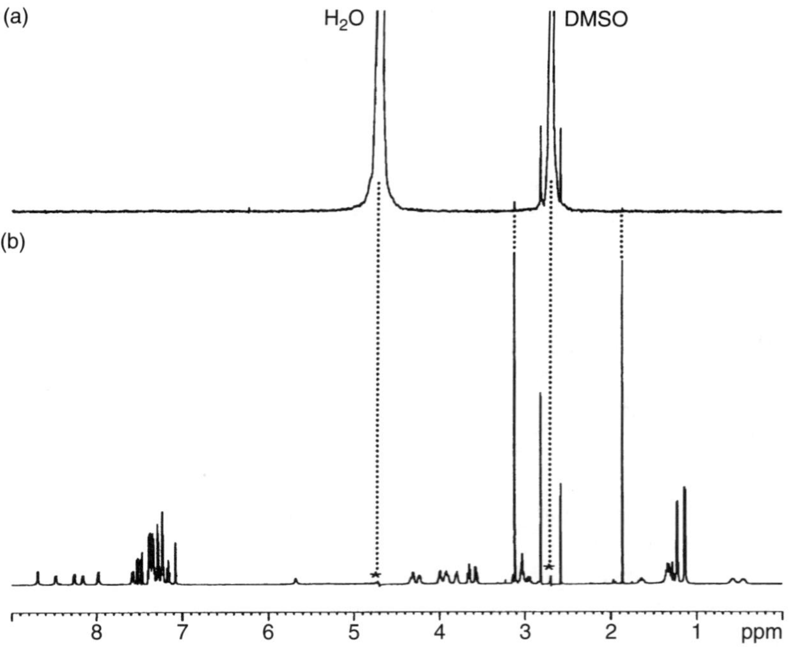

Figure 7.6 Multiple solvent suppression with the example of an NMR measurement of Sandostatin® in a 1:1 mixture of H_2O/DMSO. (The spectra were kindly made available by Novartis, Basel, Switzerland.)

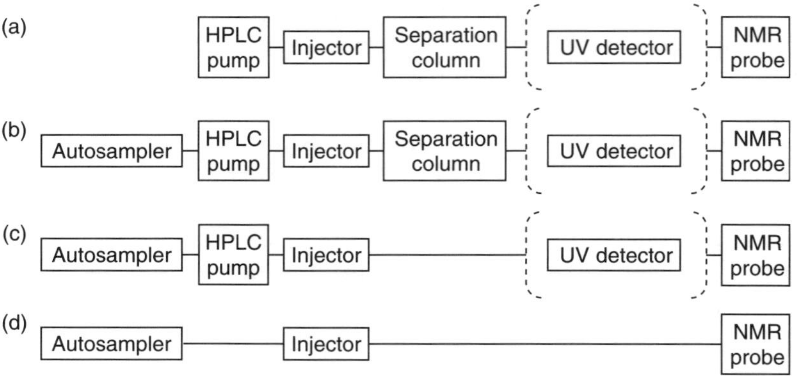

Figure 7.7 Various possibilities for HPLC–NMR coupling (a–c) or automated sample provision without separation (c and d).

arbitrarily long, for small sample quantities) [Chin *et al.*, 1998]. The coupling of HPLC and NMR puts additional demands on the suppression of the solvent signals. The solvent is constantly renewed in the probe in the flow-through mode; so-called presaturation experiments are made more difficult because of this and, furthermore, mostly solvent gradients are used, hence the resonance frequencies of the solvents are continually shifted. The suppression of these solvent signals has also been achieved with the aid of special pulse sequences [Smallcombe *et al.*, 1995].

If no compound mixtures have to be separated, but only the automatic supply of the samples to the NMR probe is of interest, the separation column can be eliminated (Figure 7.7(C)). One also speaks of flow injection in connection with this. A new spectrum could be recorded every 3 min with this setup assuming a potential measurement time of 1 min per sample and a known time span for the removal of the sample that has been measured, the cleaning of the capillaries and the supply of the next sample. This is roughly within the scope of what is required for high-throughput screening (HTS) and therefore for the solution to problem (4) [Keifer, 1998].

The system can be further simplified through elimination of the pump, detector and mobile phase (Figure 7.7(D)) by direct (undiluted) injection of the sample into the probe ($\approx$ 300 μl, $\approx$ 300 μg); measurement intervals of significantly less than 3 min, and therefore effective HTS, are made possible because of this [Keifer, 1998; Keifer *et al.*, 2000].

Problem (4) is only partially solved because of this, as the high (too high?) information content of the spectra requires time-intensive data evaluation. Despite various efforts, above all with regard to automated spectrum interpretation [Moseley and Montelione, 1999; Williams, 2000], it depends on this parameter whether, and when, fully automated real-time analysis by NMR spectroscopy will be possible within the framework of combinatorial chemistry.

7.3.3 SCREENING OF MIXTURES BY NMR

Various, very recent studies on model systems showed that NMR spectroscopy will be able to be used in the future for direct screening of compound libraries. Thus, binding affinities and structural data could be obtained simultaneously [Keifer, 1999; Shapiro and Gounarides, 1999; Siegal *et al.*, 1999].

The identification of active compounds from a library of low molecular weight organic compounds was achieved by determining the binding capabilities of the ligands to a ^{15}N-labeled receptor protein by measuring the changes of the chemical shifts of the ^{1}H or ^{15}N amide signals of the receptor protein using ^{15}N-HSQC (heteronuclear single-quantum coherence) experiments [Shuker *et al.*, 1996]. A total of 10 000 molecules were investigated; 100 measurements each day were made with 10 molecules each. Because the active compounds were able to be divided into two groups with different, but neighboring binding sites, far more active compounds were postulated based on the NMR data than if one active compound each of a group was linked via a covalent linker with an active compound of the other group. The practical experiments confirmed this assumption, which is why this approach, termed

'SAR (structure–activity relationship) with NMR' by the authors, can be considered to be a landmark for the use of NMR spectroscopy within the framework of identifying active compounds from compound mixtures.

Instead of using the change in the chemical shifts of the receptor signals as a measure for binding, the appearance of NMR signals of individual ligands can also be made use of for study of ligand–receptor interactions of a ligand mixture. The so-called pulsed field gradient (PFG) NMR diffusion technology has to be used for this, however. This permits to clearly distinguish the diffusion coefficients of a small molecule that binds to its target molecule in solution from its diffusion coefficient in the unbound state. This approach – called affinity NMR spectroscopy by the authors – allowed the differentiation of bound and unbound individual compounds directly from a small library of low molecular weight organic compounds without previous separation of the individual compounds [Lin *et al.*, 1997a] and determination of relative binding affinities of the compounds of a small library through titration of the mixture with the corresponding target molecule [Lin *et al.*, 1997b]. The stronger the interaction of a ligand with its target molecule, the lower the concentrations of the target molecule for which the ligand-specific NMR signals appear.

Furthermore, it was also able to be shown recently that the so-called transferred nuclear Overhauser effect (trNOE) can be made use of for the identification of active compounds of an oligosaccharide mixture [Meyer *et al.*, 1997]. One makes use of the fact here that, because of the binding of small molecules to high molecular weight target molecules, a heavily negative trNOE arises that distinguishes itself clearly from the NOEs of smaller, unbound molecules, which are either positive, weakly negative or not in existence at all. Further, trNOEs provide information on the three-dimensional structure of ligands, so the structural characterization of the active compound is also simultaneously possible in the bound state.

7.4 MASS SPECTROMETRY

Mass spectrometry methods are suitable both for the characterization of combinatorial product mixtures and for the analysis of small amounts of polymer-bound individual compounds in a very broad mass range. In addition, the measurements take a relatively small amount of time and can be automated, which is also applicable in part to their interpretation [Hegy *et al.*, 1996; Blom *et al.*, 1998; Wang *et al.*, 1998]. Mass spectrometry therefore represents the analysis method with the most diverse possibilities for use at the moment in the area of combinatorial chemistry [Siuzdak and Lewis, 1998; Sussmuth and Jung, 1999; Swali *et al.*, 1999; Enjalbal *et al.*, 2000], although it does not really seem to be suitable for quantitative analysis and is of a destructive nature in principle; the latter issue is not of particular importance because of the low compound quantities needed for the measurements.

7.4.1 COMMON METHODS

At the moment, four methods that are to be described in more detail below are of great significance in the analysis of combinatorial syntheses.

In electrospray ionization mass spectrometry (ESI-MS), which is schematically depicted in Figure 7.8, the sample (in solution) is applied to the instrument. The solution is guided into a metal capillary on which there is a strong electrical field. As a consequence, sample droplets that carry an excess charge are formed after leaving the capillary at atmospheric pressure. Because of the evaporation of solvent molecules, the field strength on the surface of the droplet subsequently increases to such a degree that larger droplets explode into smaller droplets and so on until the analytes enter the gas phase as singly charged ions $(M + H)^+$ (or $(M - H)^-$ in the case of measurements in the negative-ion mode) and multiply charged ions $(M + nH)^{n+}$ (or $(M - nH)^{n-}$ respectively), so they can subsequently be measured in a high vacuum. A so-called quadrupole mass filter is usually used for this purpose, which allows the selective admission of ions to the detector in accordance with their mass/charge ratio (m/z) depending on the voltage applied to the quadrupole. The use of so-called triple quadrupoles permits the specific fragmentation of ions with a certain mass (so-called tandem mass spectrometry or MS–MS); additional structural information can be obtained with the aid of the fragmentation pattern because of this. A major advantage of electrospray ionization mass spectrometry is the opportunity for direct coupling with a chromatographic method (LC–MS coupling) (Figure 7.9).

In the so-called matrix-assisted laser desorption/ionization time-of-flight (MALDI-TOF) technique, the sample is embedded in a solid matrix (for example 2,5–dihydroxybenzoic acid) (Figure 7.10). This allows for the sample molecules to be vaporized with a short laser pulse and thereby to be ionized in a relatively gentle way. Singly charged ions mainly arise in the process, which are

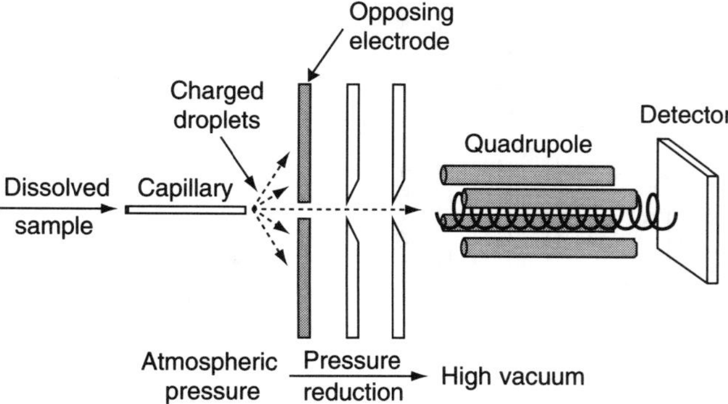

Figure 7.8 Schematic diagram of the measurement principle of electrospray mass spectrometry.

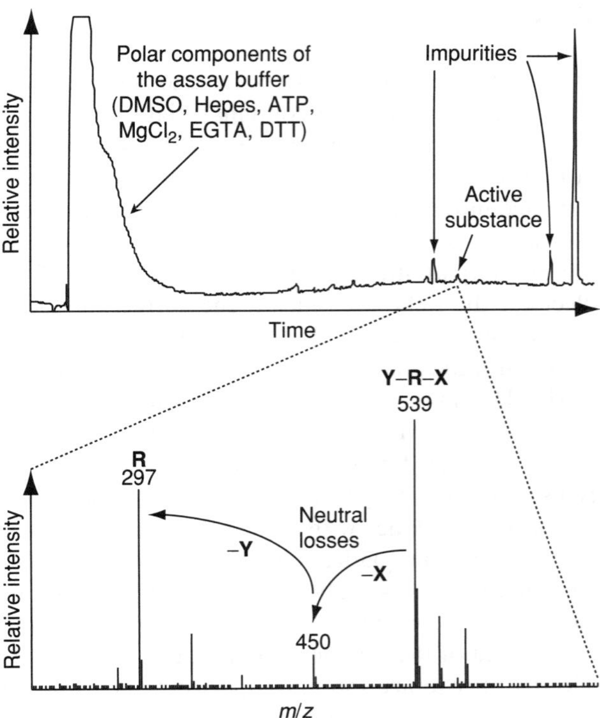

Figure 7.9 LC–MS coupling for clarifying the identity of an active compound. HPLC reveals that the active compound only represents an extremely small share of the buffer solution being tested, which makes direct mass-spectrometric analysis of the buffer impossible (above).The mass spectrum of the peak of the active compound can be selectively recorded because of the LC–MS coupling, and the identity can be determined by the fragmentation pattern (below). (The data were kindly made available by Novartis, Basel, Switzerland.)

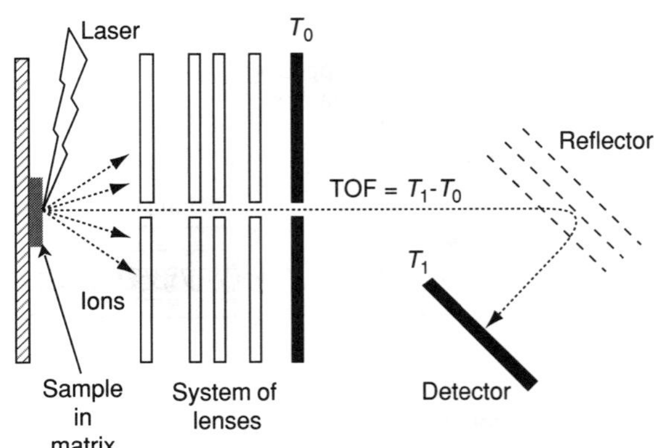

Figure 7.10 Schematic diagram of the measurement principle of MALDI-TOF mass spectrometry.

subsequently accelerated in the electric field. The mass of the ions is determined in this case by a so-called time-of-flight (TOF) analyzer, which is based on the fact that ions with the same energy but different masses require a different amount of time to traverse the same distance. The use of so-called post-source decay (PSD) technology likewise permits, analogously to the tandem mass spectrometry in the electrospray procedure, the fragmentation of ions for the generation of additional structural information.

The use of image-providing time-of-flight secondary ion mass spectrometry (TOF-SIMS) to analyze libraries provides spectra that are rather poorly interpreted in comparison to the methods described previously [Brummel *et al.*, 1996] and is not in very widespread use at the moment. With this technique it is possible to scan sample surfaces or beads, however, and to therefore work with a high spatial resolution, hence this method offers a great potential for development. The schematic diagram of TOF-SIMS mass spectrometry essentially corresponds to Figure 7.10 (even if the actual setup of the instrument is different); in contrast, the samples can be directly measured on the polymer because the samples are ionized with a gallium ion beam (Ga^+) instead of with a laser. The surface of the sample can be mapped with its help under formation of the secondary ions with a spatial resolution in the upper nanometer range; a two-dimensional image of the sample can be created because of this.

Fourier-transform ion cyclotron resonance mass spectrometers (FT-ICR-MS) are probably the most complex mass spectrometers, but they offer an incredible mass resolution. They consist of three sections: a sample source (ESI and MALDI are the most common), the ion transfer region, where the ions are focussed, bunched and transferred into the high vacuum, and the analyzer itself. Typically, the analyser consists of an ion-trap located within a spatially uniform static magnetic field, which constrains the incident ions in a circular orbit. The ions of a given m/z are excited to a larger orbital radius by applying a radio frequency sweep across the cell. One frequency excites one particular m/z, but Fourier transformation allows the simultaneous excitation of all frequencies. Measurement of the angular frequency leads to values for m/z and thus to the mass spectrum. The specifically selective ion trapping, the sensitivity, the high resolution and mass accuracy over a broad mass range make FT-ICR-MS one of the most promising techniques for the analysis of combinatorial libraries, even without prior chromatographic separation [Walk *et al.*, 1999; Poulsen *et al.*, 2000].

7.4.2 SEQUENCING OF PEPTIDES BY MEANS OF MS

The sequencing of peptides with ESI, MALDI or FT-ICR mass spectrometry represents a real alternative to the sequencing of peptides by means of Edman degradation. Although only individual peptides can be sequenced because a pool sequencing of mixtures, as is possible with the Edman degradation, is ruled out by this approach (see section on microsequencing of peptide libraries in Chapter 4).

The selective fragmentation of ions by means of MS–MS or PSD which has already been described in the previous section is used for sequencing; it is particularly useful in connection with this that a peptide bond can be broken more simply in general than the other bonds of the peptide backbone or the side chains (Figure 7.11). The resulting fragmentation pattern can subsequently be interpreted manually or with the aid of appropriate software, and the sequence can be deduced from this. The structural isomers leucine and isoleucine, represent a problem because they have identical mass. Youngquist *et al.* have developed a MALDI method for synthetic peptides, however, with which they are able to quickly sequence peptides and simultaneously solve the leucine/isoleucine problem in an elegant way [Youngquist *et al.*, 1995]. A small

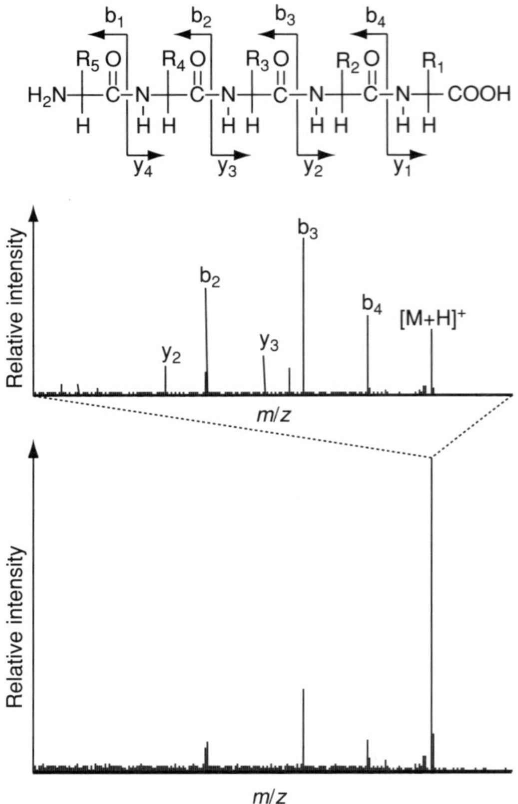

Figure 7.11 Electrospray mass spectrum of a pentapeptide (below) with fragmentation of the quasi molecular ion $[M + H]^+$ via tandem mass spectrometry (center); the sequence of the peptide can be derived because of this (above).The nomenclature of the fragments corresponds to the customary designation rules.

percentage of a capping reagent is added after every coupling step for this, which leads in each case to a sequence-specific chain-termination product. As a result, a complete series of the chain-termination products, from which the sequence can be directly read without fragmentation experiments, also appears in the mass spectrum in addition to the mass of the final product. Amino acids which have a partner with the same or a similar mass are 'capped' with different reagents or reagent mixtures, thus it is subsequently possible to easily identify them. This method is also called ladder sequencing.

7.4.3 QUALITY CONTROL OF LIBRARIES BY MS

It was able to be shown for the first time with the example of investigation of peptide libraries by ESI-MS that mass spectrometry methods are very well suited for checking the quality of libraries [Metzger *et al.*, 1993; Stefanovic *et al.*, 1993; Metzger *et al.*, 1994]. Starting with the synthesis design of a library, precise composition of the peptides and consequently distribution of the masses to be expected can be calculated. The quality of the library can then be estimated quite well through a comparison of the theoretical mass distribution with the mass spectrum of the complete library that is actually measured (Figure 7.12). A failed coupling that involves every peptide of the library leads to a shift in the calculated pattern to a lower m/z ratio; a partial modification that involves all of the peptides of the library can be recognized by the appearance of a second mass pattern having an identical distribution with a greater m/z ratio. Side reactions in the synthesis of peptide libraries can consequently also be identified and characterized. If the calculated mass distribution has no similarity at all to the measured mass spectrum, however, it is extremely difficult to provide an explanation for the poor quality of the library with only the aid of the mass spectrum.

In the meanwhile, the quality control of libraries by mass spectrometry has also been applied to a series of nonpeptide libraries, such as a mixture of modified xanthene derivatives, for instance [Dunayevskiy *et al.*, 1995].

The verification that every component actually exists in a library, that all of the library members are distributed in a roughly equal way or which library member could possibly not exist can only be provided with great difficulty by means of mass spectrometry. If at all, this only succeeds in smaller libraries (up to around 100 library members) through use of the fragmentation techniques or on-line coupling with chromatographic methods mentioned in the previous sections, as was able to be shown in the example, among others, of the MS analysis of a diketopiperazine library [Gordon and Steele, 1995].

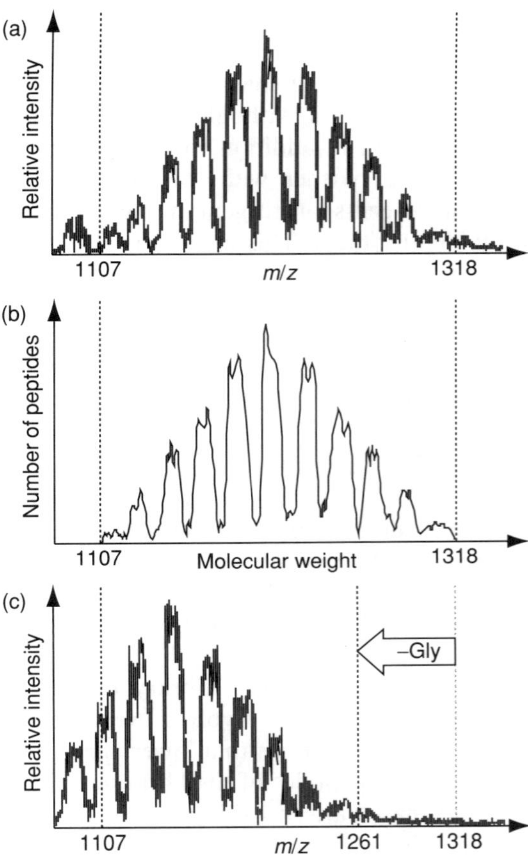

Figure 7.12 (a) Electrospray mass spectrum of a synthetic peptide library that consists of several thousand individual compounds (region of the $[M + H]^+$ ions); the vertical dotted lines identify the calculated masses of the lightest and of the heaviest peptides of the library. (b) Theoretical mass distribution of the peptide library, i.e. the calculated number of peptides of the same mass with reference to the molecular weight of their $[M + H]^+$ ions. (c) Electrospray mass spectrum of the identical peptide library with a completely failed coupling of a defined glycine (following Metzger *et al.* [1993] and Metzger *et al.* [1994]).

7.4.4 MASS SPECTROMETRY OF INDIVIDUAL BEADS

The mass spectrometry analysis of peptides and other compounds on individual beads is possible on the grounds of the high sensitivity of mass spectrometers, which permits the detection of sample quantities in the femto to attomolar range, by means of ESI-MS, MALDI-TOF-MS, TOF-SIMS and FT-ICR-MS (Figure 7.13). Only separation of the immobilized products from the respective polymeric support proves to be a problem. Acid-labile linkers are mainly used to solve this problem, so that the bond of the product to the bead

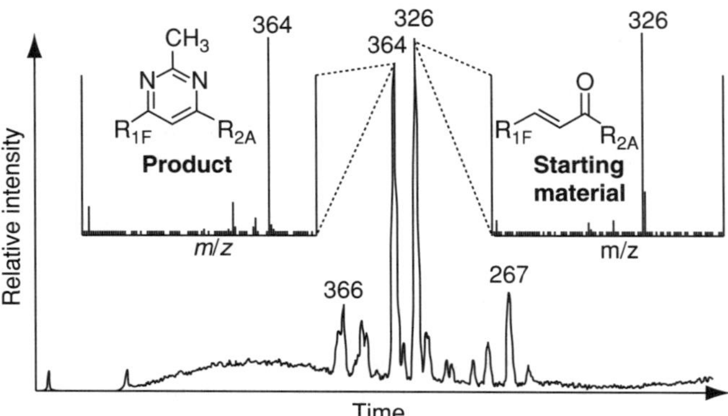

Figure 7.13 LC–MS base peak chromatogram of the compounds bound to a single bead with electrospray mass spectra of the prominent signals. The products were cleaved from the polymer in advance with trifluoroacetic acid (TFA) and separated by means of on-line HPLC coupling. It is apparent that the starting material (right) was only able to be incompletely converted into the desired product (left). (The data were kindly made available by Novartis, Basel, Switzerland.)

can be cleaved by treatment with gaseous trifluoroacetic acid. Depending on the method, the product can then be directly measured (TOF-SIMS) [Brummel *et al.*, 1994] or has to be dissolved (ESI-MS) or embedded in a matrix (MALDI-TOF-MS) [Zambias *et al.*, 1994; Egner *et al.*, 1995].

Brown *et al.* used a photolabile linker; they were able to avoid complicated work with trifluoroacetic acid and as a result separated the product from the polymer via irradiation with a UV lamp [Brown *et al.*, 1995]. But the product had to be further processed, i.e. dissolved, for their measurements by means of ESI-MS. Fitzgerald *et al.* [1996] introduced a very elegant alternative. They used a MALDI-TOF instrument with a laser which permitted the bead-bound product embedded in a matrix to be separated from it, and then vaporized and ionized in a single step.

REFERENCES

Albert, K. (1995). Direct online coupling of capillary electrophoresis and 1H NMR spectroscopy. *Angew. Chem. Int. Ed. Engl.* **34**, 641–642.

Albert, K., Braumann, U., Tseng, L. H. and Schlotterbeck, G. (1998). On-line coupling of capillary HPLC with 1H NMR spectroscopy. *Biomed. Chromatogr.* **12**, 158–159.

Barn, D. R., Morphy, J. R. and Rees, D. C. (1996). Synthesis of an array of amides by aluminium chloride assisted cleavage of resin-bound esters. *Tetrahedron Lett.* **37**, 3213–3216.

Berger, T. A., Fogleman, K., Staats, T., Bente, P., Crocket, I., Farrell, W. and Osonubi, M. (2000). The development of a semi-preparatory scale supercritical-fluid chromatograph for high-throughput purification of 'combi-chem' libraries. *J. Biochem. Biophys. Methods*, **43**, 87–111.

Berger, T. A. and Wilson, W. H. (2000). High-speed screening of combinatorial libraries by gradient packed-column supercritical fluid chromatography. *J. Biochem. Biophys. Methods*, **43**, 77–85.

Blom, K. F., Combs, A. P., Rockwell, A. L., Oldenburg, K. R., Zhang, J. -H. and Chen, T. (1998). Direct mass spectrometric determination of bead bound compounds in a combinatorial lead discovery application. *Rapid Commun. Mass Spectrom.* **12**, 1192–1198.

Briceno, G., Chang, H. Y., Sun, X. D., Schultz, P. G. and Xiang, X. D. (1995). A class of cobalt oxide magnetoresistance materials discovered with combinatorial synthesis. *Science*, **270**, 273–275.

Brown, B. B., Wagner, D. S. and Geysen, H. M. (1995). A single-bead decode strategy using electrospray ionization mass spectrometry and a new photolabile linker: 3-amino-3-(2-nitrophenyl) propionic acid. *Mol. Divers.* 1, pp. 4–12.

Brummel, C. L., Lee, I. N., Zhou, Y., Benkovic, S. J. and Winograd, N. (1994). A mass spectrometric solution to the address problem of combinatorial libraries. *Science*, **264**, 399–402.

Brummel, C. L., Vickerman, J. C., Carr, S. A., Hemling, M. E., Roberts, G. D., Johnson, W., Weinstock, J., Gaitanopoulos, D., Benkovic, S. J. and Winograd, N. (1996). Evaluation of mass spectrometric methods applicable to the direct analysis of non-peptide bead-bound combinatorial libraries. *Anal. Chem.* **68**, 237–242.

Chin, J., Fell, J. B., Jarosinski, M., Shapiro, M. J. and Wareing, J. R. (1998). HPLC/NMR in combinatorial chemistry. *J. Org. Chem.* **63**, 386–390.

Coleman, K. (1999). High-throughput preparative separations from combinatorial libraries. *Analysis*, **27**, 719–723.

DeWitt, S. H., Kiely, J. S., Stankovic, C. J., Schroeder, M. C., Cody, D. M. and Pavia, M. R. (1993). 'Diversomers': an approach to nonpeptide, nonoligomeric chemical diversity. *Proc. Natl. Acad. Sci. USA*, **90**, 6909–6913.

Doskocilova, D., Dang, D. T. and Schneider, B. (1970). Narrowing of proton NMR lines by magic angle rotation. *Chem. Phys. Lett.* **6**, 381–384.

Doskocilova, D., Schneider, B. and Trekoval, J. (1974). Characterization of internal motions in crosslinked polymer gels by high-resolution MAS NMR spectrometry. *Collect. Czech. Chem. Commun.* **39**, 2943–2948.

Doskocilova, D., Schneider, B. and Jakes, J. (1978). NMR spectra of systems with restricted motion: cross-linked polymer gels. *J. Magn. Reson.* **29**, 79–90.

Dunayevskiy, Y., Vouros, P., Carell, T., Wintner, E. A. and Rebek, J. (1995). Characterization of the complexity of small-molecule libraries of electrospray ionization mass spectrometry. *Anal. Chem.* **67**, 2906–2915.

Egner, B. J. and Bradley, M. (1997). Analytical techniques for solid-phase organic and combinatorial synthesis. *Drug Discov. Today*, **2**, 102–109.

Egner, B. J., Langley, G. J. and Bradley, M. (1995). Solid-phase chemistry – direct monitoring by matrix-assisted laser-desorption ionization time-of-flight mass-spectrometry – a tool for combinatorial chemistry. *J. Org. Chem.* **60**, 2652–2653.

Enjalbal, C., Martinez, J. and Aubagnac, J. L. (2000). Mass spectrometry in combinatorial chemistry. *Mass Spectrom. Rev.* **19**, 139–161.

Fang, L., Wan, M., Pennacchio, M. and Pan, J. (2000). Evaluation of evaporative light-scattering detector for combinatorial library quantitation by reversed phase HPLC. *J. Comb. Chem.* **2**, 254–257.

Fitch, W. L., Detre, G., Holmes, C. P., Shoolery, J. N. and Keifer, P. A. (1994). High-resolution 1H NMR in solid-phase organic synthesis. *J. Org. Chem.* **59**, 7955–7956.

Fitzgerald, M. C., Harris, K., Shevlin, C. G. and Siuzdak, G. (1996). Direct characterization of solid phase resin-bound molecules by mass spectrometry. *Bioorg. Med. Chem. Lett.* **6**, 979–982.

Forman, F. W. and Sucholeiki, I. (1995). Solid-phase synthesis of biaryls via the Stille reaction. *J. Org. Chem.* **60**, 523–528.

Frechet, J. M. and Schuerch, C. (1971). Solid-phase synthesis of oligosaccharides. I. Preparation of the solid support poly(*p*-(1-propen-3-ol-1-yl)styrene). *J. Am. Chem. Soc.* **93**, 492–496.

Gallop, M. A. and Fitch, W. L. (1997). New methods for analyzing compounds on polymeric supports. *Curr. Opin. Chem. Biol.* **1**, 94–100.

Gaus, H. J., Kung, P. -P., Brooks, D., Cook, P. D. and Cummins, L. L. (1999). Monitoring solution-phase combinatorial library synthesis by capillary electrophoresis. *Biotechnol. Bioeng.* **61**, 169–177.

Gordeev, M. F., Patel, D. V. and Gordon, E. M. (1996). Approaches to combinatorial synthesis of heterocycles – a solid-phase synthesis of 1,4-dihydropyridines. *J. Org. Chem.* **61**, 924–928.

Gordon, D. W. and Steele, J. (1995). Reductive alkylation on a solid phase: synthesis of a piperazinedione combinatorial library. *Bioorg. Med. Chem. Lett.* **5**, 47–50.

Gremlich, H.-U. (1999). The use of optical spectroscopy in combinatorial chemistry. *Biotechnol. Bioeng.* **61**, 179–187.

Griffey, R. H., An, H., Cummins, L. L., Gaus, H. J., Haly, B., Herrmann, R. and Dan Cook, P. (1998). Rapid deconvolution of combinatorial libraries using HPLC fractionation. *Tetrahedron*, **54**, 4067–4076.

Haap, W. J., Kaiser, D., Walk, T. B. and Jung, G. (1998). Solid phase synthesis of diverse isoxazolidines via 1,3–dipolar cycloaddition. *Tetrahedron*, **54**, 3705–3724.

Han, H. and Janda, K. D. (1996). Azatides: solution and liquid phase syntheses of a new peptidomimetic. *J. Am. Chem. Soc.* **118**, 2539–2544.

Han, H., Wolfe, M. M., Brenner, S. and Janda, K. D. (1995). Liquid-phase combinatorial synthesis. *Proc. Natl. Acad. Sci. USA*, **92**, 6419–6423.

Hegy, G., Gorlach, E., Richmond, R. and Bitsch, F. (1996). High throughput electrospray mass spectrometry of combinatorial chemistry racks with automated contamination surveillance and results reporting. *Rapid Commun. Mass Spectrom.* **10**, 1894–1900.

Hsu, B. H., Orton, E., Tang, S. -Y. and Carlton, R. A. (1999). Application of evaporative light scattering detection to the characterization of combinatorial and parallel synthesis libraries for pharmaceutical drug discovery. *J. Chromatogr., B*, **725**, 103–112.

Johnson, C. R. and Zhang, B. (1995). Solid phase synthesis of alkenes using the Horner–Wadsworth–Emmons reaction and monitoring by gel phase (31)P NMR. *Tetrahedron Lett.* **36**, 9253–9256.

Keifer, P. A. (1996). Influence of resin structure, tether length, and solvent upon the high resolution 1H NMR in solid-phase organic synthesis. *J. Org. Chem.* **61**, 1558–1559.

Keifer, P. A. (1998). New methods for obtaining high-resolution NMR spectra of solid-phase synthesis resins, natural products, and solution-state combinatorial chemistry libraries. *Drugs Future*, **23**, 301–317.

Keifer, P. A. (1999). NMR tools for biotechnology. *Curr. Opin. Biotechnol.* **10**, 34–41.

Keifer, P. A., Smallcombe, S. H., Williams, E. H., Salomon, K. E., Mendez, G., Belletire, J. L. and Moore, C. D. (2000). Direct-injection NMR (DI-NMR): a flow NMR technique for the analysis of combinatorial chemistry libraries. *J. Comb. Chem.* **2**, 151–171.

Kempe, M., Keifer, P. A. and Barany, G. (1997). CLEAR supports for solid-phase synthesis. In: Ramage, R. and Epton, R. (eds), *Peptides 1996 (Proceedings of the 24th European Peptide Symposium)*, Mayflower Scientific Ltd, Kingswinford, UK, pp. 533–534.

Kibbey, C. E. (1996). Quantitation of combinatorial libraries of small organic molecules by normal-phase HPLC with evaporative light-scattering detection. *Mol. Divers.* **1**, 247–258.

Kim, R. M., Manna, M., Hutchins, S. M., Griffin, P. R., Yates, N. A., Bernick, A. M. and Chapman, K. T. (1996). Dendrimer-supported combinatorial chemistry. *Proc. Natl. Acad. Sci. USA*, **93**, 10 012–10 017.

Kirkland, J. J. (2000). Ultrafast reversed-phase high-performance liquid chromatographic separations: an overview. *J. Chromatogr. Sci.* **38**, 535–544.

Korhammer, S. A. and Bernruether, A. (1996). Hyphenation of high-performance liquid chromatography (HPLC) and other chromatographic techniques (SFC, GPC, GC, CE) with nuclear magnetic resonance (NMR). *Fresenius J. Anal. Chem.* **354**, 131–135.

Leibfritz, D., Mayr, W., Oekonomopulos, R. and Jung, G. (1978). Carbon-13 NMR spectroscopic studies on the conformation during stepwise synthesis of peptides bound to solubilizing polymer supports. *Tetrahedron*, **34**, 2045–2050.

Lewis, K., Phelps, D. and Sefler, A. (2000). Automated high-throughput quantification of combinatorial arrays. *Am. Pharm. Rev.* **3**, 63–68.

Li, W. B. and Yan, B. (1998). Effects of polymer supports on the kinetics of solid-phase organic reactions – a comparison of polystyrene- and TentaGel-based resins. *J. Org. Chem.* **63**, 4092–4097.

Lin, M. F., Shapiro, M. J. and Wareing, J. R. (1997a). Screening mixtures by affinity NMR. *J. Org. Chem.* **62**, 8930–8931.

Lin, M. F., Shapiro, M. J. and Wareing, J. R. (1997b). Diffusion-edited NMR – affinity NMR for direct observation of molecular interactions. *J. Am. Chem. Soc.* **119**, 5249–5250.

Look, G. C., Holmes, C. P., Chinn, J. P. and Gallop, M. A. (1994). Methods for combinatorial organic synthesis: the use of fast 13C NMR analysis for gel phase reaction monitoring. *J. Org. Chem.* **59**, 7588–7590.

Ma, L., Gong, X. and Yeung, E. S. (2000). Combinatorial screening of enzyme activity by using multiplexed capillary electrophoresis. *Anal. Chem.* **72**, 3383–3387.

Manatt, S. L., Amsden, C. F., Bettison, C. A., Frazer, W. T., Gudman, J. T., Lenk, B. E., Lubetich, J. F., McNelly, E. A., Smith, S. C., Templeton, D. J. and Pinnell, R. P. (1980a). A fluorine-19 NMR approach for studying Merrifield solid-phase peptide syntheses. *Tetrahedron Lett.* **21**, 1397–1400.

Manatt, S. L., Horowitz, D., Horowitz, R. and Pinnell, R. P. (1980b). Solvent swelling for enhancement of carbon-13 nuclear magnetic resonance spectral information from insoluble polymers: chloromethylation levels in cross-linked polystyrenes. *Anal. Chem.* **52**, 1529–1532.

Metzger, J. W., Kempter, C., Wiesmueller, K. H. and Jung, G. (1994). Electrospray mass spectrometry and tandem mass spectrometry of synthetic multicomponent peptide mixtures: determination of composition and purity. *Anal. Biochem.* **219**, 261–277.

Metzger, J. W., Wiesmueller, K. H., Gnau, V., Bruenjes, J. and Jung, G. (1993). Ion-spray mass spectrometry and high-performance liquid chromatography. Mass spectrometry of synthetic peptide libraries. *Angew. Chem. Int. Ed. Engl.* **32**, 894–896.

Meyer, B., Weimar, T. and Peters, T. (1997). Screening mixtures for biological activity by NMR. *Eur. J. Biochem.* **246**, 705–709.

Moseley, H. N. B. and Montelione, G. T. (1999). Automated analysis of NMR assignments and structures for proteins. *Curr. Opin. Struct. Biol.* **9**, 635–642.

Olson, D. L., Peck, T. L., Webb, A. G. and Sweedler, J. V. (1996). On-line NMR detection for capillary electrophoresis applied to peptide analysis, 730–731. In: *Peptides: Chemistry, Structure and Biology (Proceedings of the 14th American Peptide Symposium)* (Edited by P. T. P. Kaumaya and R. S. Hodges), Mayflower Scientific Ltd, Kingswinford, UK.

Peng, S. X. (2000). Hyphenated HPLC–NMR and its applications in drug discovery. *Biomed. Chromatogr.* **14**, 430–441.

Poulsen, S.-A., Gates, P. J., Cousins, G. R. L. and Sanders, J. K. M. (2000). Electrospray ionisation Fourier-transform ion cyclotron resonance mass spectrometry of dynamic combinatorial libraries. *Rapid Commun. Mass Spectrom.* **14**, 44–48.

Pusecker, K., Schewitz, J., Gfrorer, P., Tseng, L. H., Albert, K. and Bayer, E. (1998). On line coupling of capillary electrochromatography, capillary electrophoresis, and capillary HPLC with nuclear magnetic resonance spectroscopy. *Anal. Chem.* **70**, 3280–3285.

Sarkar, S. K., Garigipati, R. S., Adams, J. L. and Keifer, P. A. (1996). An NMR method to identify nondestructively chemical compounds bound to a single solid-phase-synthesis bead for combinatorial chemistry applications. *J. Am. Chem. Soc.* **118**, 2305–2306.

Schaefer, J. (1971). High-resolution pulsed carbon-13 nuclear magnetic resonance analysis of some cross-linked polymers. *Macromolecules*, **4**, 110–112.

Shah, N., Gao, M., Tsutsui, K., Lu, A., Davis, J., Scheuerman, R., Fitch, W. L. and Wilgus, R. L. (2000). A novel approach to high-throughput quality control of parallel synthesis libraries. *J. Comb. Chem.* **2**, 453–460.

Shapiro, M. J. and Gounarides, J. S. (1999). NMR methods utilized in combinatorial chemistry research. *Prog. Nucl. Magn. Reson. Spectrosc.* **35**, 153–200.

Shapiro, M. J., Kumaravel, G., Petter, R. C. and Beveridge, R. (1996). 19F NMR monitoring of a S(N)Ar reaction on solid support. *Tetrahedron Lett.* **37**, 4671–4674.

Shuker, S. B., Hajduk, P. J., Meadows, R. P. and Fesik, S. W. (1996). Discovering high-affinity ligands for proteins: SAR by NMR. *Science*, **274**, 1531–1534.

Sidelmann, U. G., Gavaghan, C., Carless, H. A. J., Spraul, M., Hofmann, M., Lindon, J. C., Wilson, I. D. and Nicholson, J. K. (1995). 750-MHz directly coupled HPLC–NMR: application for the sequential characterization of the positional isomers and anomers of 2–, 3–, and 4–fluorobenzoic acid glucuronides in equilibrium mixtures. *Anal. Chem.* **67**, 4441–4445.

Siegal, G., Van Duynhoven, J. and Baldus, M. (1999). Biomolecular NMR: recent advances in liquids, solids and screening. *Curr. Opin. Chem. Biol.* **3**, 530–536.

Siuzdak, G. and Lewis, J. K. (1998). Applications of mass spectrometry in combinatorial chemistry. *Biotechnol. Bioeng.* **61**, 127–134.

Smallcombe, S. H., Patt, S. L. and Keifer, P. A. (1995). WET solvent suppression and its applications to LC NMR and high-resolution NMR spectroscopy. *J. Magn. Reson.* **117**, 295–303.

Somsen, G. W., Gooijer, C., Velthorst, N. H. and Brinkman, U. A. T. (1998). Coupling of column liquid chromatography and Fourier transform infrared spectrometry. *J. Chromatogr.* **811**, 1–34.

Stefanovic, S., Wiesmueller, K. H., Metzger, J. W., Beck-Sickinger, A. G. and Jung, G. (1993). Natural and synthetic peptide pools: characterization by sequencing and electrospray mass spectrometry. *Bioorg. Med. Chem. Lett.* **3**, 431–436.

Stockman, B. J. (2000). Flow NMR spectroscopy in drug discovery. *Curr. Opin. Drug Discov. Dev.* **3**, 269–274.

Stoever, H. D. H. and Frechet, J. M. J. (1989). Direct polarization 13C and 1H magic angle spinning NMR in the characterization of solvent-swollen gels. *Macromolecules*, **22**, 1574–1576.

Strohschein, S., Schlotterbeck, G., Richter, J., Pursh, M., Tseng, L. H., Haendel, H. and Albert, K. (1997). Comparison of the separation of *cis/trans* isomers of tretinoin with different stationary phases by liquid chromatography–nuclear magnetic resonance coupling. *J. Chromatogr.* **765**, 207–214.

Sussmuth, R. D. and Jung, G. (1999). Impact of mass spectrometry on combinatorial chemistry. *J. Chromatogr., B*, **725**, 49–65.

Swali, V., Langley, G. J. and Bradley, M. (1999). Mass spectrometric analysis in combinatorial chemistry. *Curr. Opin. Chem. Biol.* **3**, 337–341.

Taylor, E. W., Qian, M. G. and Dollinger, G. D. (1998). Simultaneous online characterization of small organic molecules derived from combinatorial libraries for identity, quantity, and purity by reversed-phase HPLC with chemiluminescent nitrogen, UV, and mass spectrometric detection. *Anal. Chem.* **70**, 3339–3347.

Walk, T. B., Trautwein, A. W., Richter, H. and Jung, G. (1999). ESI Fourier transform ion cyclotron resonance mass spectrometry (ESI-FT-ICR-MS): a rapid high-resolution analytical method for combinatorial compound libraries. *Angew. Chem. Int. Ed. Engl.* **38**, 1763–1765.

Wang, T., Zeng, L., Strader, T., Burton, L. and Kassel, D. B. (1998). A new ultra-high throughput method for characterizing combinatorial libraries incorporating a multiple probe autosampler coupled with flow injection mass spectrometry analysis. *Rapid Commun. Mass Spectrom.* **12**, 1123–1129.

Williams, A. (2000). Recent advances in NMR prediction and automated structure elucidation software. *Curr. Opin. Drug Discov. Dev.* **3**, 298–305.

Wu, N., Peck, T. L., Webb, A. G., Magin, R. L. and Sweedler, J. V. (1994). 1H-NMR spectroscopy on the nanoliter scale for static and on-line measurements. *Anal. Chem.* **66**, 3849–3857.

Yan, B., Fell, J. B. and Kumaravel, G. (1996a). Progression of organic reactions on resin supports monitored by single bead FTIR microspectroscopy. *J. Org. Chem.* **61**, 7467–7472.

Yan, B. and Kumaravel, G. (1996). Probing solid-phase reactions by monitoring the IR bands of compounds on a single 'flattened' resin bead. *Tetrahedron*, **52**, 843–848.

Yan, B., Kumaravel, G., Anjaria, H., Wu, A., Petter, R. C., Jewell, C. F. and Wareing, J. R. (1995). Infrared spectrum of a single resin bead for real-time monitoring of solid-phase reactions. *J. Org. Chem.* **60**, 5736–5738.

Yan, B., Sun, Q., Wareing, J. R. and Jewell, C. F. (1996b). Real-time monitoring of the catalytic oxidation of alcohols to aldehydes and ketones on resin support by single-bead Fourier transform infrared microspectroscopy. *J. Org. Chem.* **61**, 8765–8770.

Youngquist, R. S., Fuentes, G. R., Lacey, M. P. and Keough, T. (1995). Generation and screening of combinatorial peptide libraries designed for rapid sequencing by mass-spectrometry. *J. Am. Chem. Soc.* **117**, 3900–3906.

Zambias, R. A., Boulton, D. A. and Griffin, P. R. (1994). Microchemical structural determination of a peptoid covalently bound to a polymeric bead by matrix-assisted laser desorption ionization time-of-flight mass spectrometry. *Tetrahedron Lett.* **35**, 4283–4286.

Glossary

Activity Measure of strength of the effect of a compound to be tested. The compound can be tested in the test tube with isolated target molecules (*in vitro*), on cells that contain the target molecule, or in an animal (*in vivo*).

Affinity Measure of strength of the bond of a compound to be tested to a biological target molecule, for example the binding of a hormone derivative to a receptor.

Anchor Chemical group that creates the direct connection between the product and the polymeric support. The connection has to remain stable during the entire solid phase synthesis and only be able to be cleaved at the end of the synthesis, with the formation of the desired functional group on the product.

Aptamer RNA molecule (around 40–200 bases) that can bind a ligand. Aptamers can frequently be used in a way similar to antibodies, but are not labile towards proteases and can be optimized *in vitro*.

Array Designation for a parallel library. The products usually take on a special format, for example chessboard style on a chip surface or in the wells of a microtiter plate.

Assay Experiment in which the properties of compounds are tested.

Bacteriophages Viruses that infect bacteria, for example *Escherichia coli (E. coli)*.

Bead Porous polymer granule (made of cross-linked polystyrene, for example) equipped with functional groups for the synthesis of libraries, in particular according to the 'one bead one compound' concept.

Binary encoding Encoding technique of a library that is based on the presence or absence of tags. Analogously to the fundamentals of computer technology ('current flows/current does not flow'), a binary sequence of numbers (for example '0101011') arises because of this that represents, and unambiguously describes, the progress of a certain synthesis process.

Bioavailability Describes the amount of drug that is absorbed or becomes available at the site of physiological activity after administration.

Biopolymers Natural compounds such as proteins or carbohydrates that are constructed out of a multitude of identical or similar building blocks. The linkage can take place in a linear or in a branched fashion.

Capping Blocking of unconverted functional groups through highly reactive compounds, in order to prevent the creation of undesired byproducts in the further progress of the synthesis process.

Compound pool Internal company collection of all compounds that have ever been produced in a sufficient quantity. The compound pool is used in the search for lead compounds, just as the natural substance pools and libraries are used, in order to obtain compounds with 'interesting' properties as quickly as possible.

Coupling Formation of a covalent bond through reaction of the functional group of a chemical compound with that of a reaction partner. One of the reaction partners is usually polymer bound.

Deconvolution Derivation of the identity of an individual compound with the desired properties from the totality of the compounds of a library.

'Defined' position Sequence position (usually of a peptide) at which several amino acids, for example the 20 that occur naturally, are coupled in separate, parallel sets during the synthesis. As a result, if a 'mixed' position also exists in the sequence, the corresponding number of sublibraries arise.

DNA replication Doubling of the DNA, necessary in cell division.

ELISA Enzyme-linked immunosorbent assay. Enzymatic test that is based on the specific reaction of antibodies with their antigen. The recognition takes place through a second antibody that has been directed, for example, at the constant part of the first one and that was coupled to an enzyme, which catalyzes a color reaction, for instance.

Encoding If no analysis system has been established for the members of a library with which a conclusion can be made as to the identity of the 'most interesting' compounds in the subsequent structure determination, an attempt is made through the introduction of tags to make this information accessible.

Epitope Area of a protein that is recognized by an antibody.

Extraction Method for purification of compound mixtures, for example after synthesis. It is frequently used in combinatorial chemistry, because it can be automated. A distinction is made between liquid–liquid extraction and solid phase extraction. In liquid–liquid extraction, the different distribution of the compounds to be separated in aqueous and organic solvents that do not mix and that form two phases is made use of. In solid phase extraction, the compounds to be separated are distinguished by their binding ability to a solid phase.

Freeze-drying Also called lyophilization. Gentle-action method to remove water, for instance. Sensitive, aqueous solutions of peptides or proteins are frozen, and the solvent is sublimed by high vacuum.

Green fluorescent protein (GFP) Protein that was isolated from jellyfish and that has its own fluorescence. It can be produced in a recombinant fashion and used in cells as an internal label.

Heteroatom substitution Replacement of CH_2 by O, NH, or S which results in compounds with new properties, for example with higher polarity.

High-throughput screening (HTS) Testing of a very high number of compounds, as frequently result in the synthesis of libraries, in a short period of time, i.e. with a high throughput. The use of computer-controlled robot systems is indispensable for this in general.

Lead compound First compound in the development of an active substance that has the desired properties, but has to be further optimized, for example a screened compound with a low affinity but a certain receptor selectivity.

Library A library includes a multitude of very diverse compounds. Explicitly knowing the identity of every single one of these compounds is not of interest initially. This multitude is reduced to molecules with the 'most interesting' properties by the obligatory screening, and one only then attempts to determine their identity.

Linker Chemical group that connects one molecule with another. The linker connects the product to be synthesized with the polymeric support within the framework of solid phase synthesis.

Lyophilization Freeze-drying.

Mapping Analysis of the sequence of a protein with regard to a desired property, for example recognition by an antibody. This is typically done by synthesizing the sequence of the protein in the form of small, overlapping peptides, which are then individually tested for the desired property.

Mimetics Compounds that share the positive properties of other molecules, for example a certain affinity for a receptor, on the grounds of certain similarities. But they show no relation with regard to other, negative properties, for instance a low protease stability.

Mimotope Compound that imitates the epitope of an antibody and is therefore likewise recognized by the antibody.

'Mixed' position Sequence position (usually of a peptide) at which no certain amino acid, but instead several, for example a mixture of the 20 natural amino acids, are coupled during synthesis. As a result, a library forms.

Mutant, mutation Analogue of a protein with the exchange of one or more amino acids. Natural mutations are distinguished from generated ones. The latter serve to identify relevant amino acids in proteins.

'One bead one compound' Strategy for the synthesis of libraries in which each polymer bead only carries a single kind of compound in each case as a result.

Orthogonal Designation for the use of protecting groups that are based on a reciprocal cleavage concept in a single synthesis strategy. The base lability of the *N*-terminal protecting group and the acid lability of the side-chain protecting groups are an example within the framework of the 9-fluorenylmethoxycarbonyl (Fmoc) peptide synthesis strategy.

Parallel synthesis Simultaneous, spatially separated synthesis of several compounds in a synthesis set.

Peptoids Oligoglycines that are substituted at the nitrogen atoms of the peptide bonds.

Phage display Technique for creating relevant peptides or proteins on phage proteins with the aid of randomly generated, biological libraries and for screening with a suitable test system.

Phagemid Plasmid that is packed in phages and can therefore be amplified.

Pin Small rod that is supplied with a functionalized polymer at one end, in order to carry out solid phase syntheses on it. The arrangement of 96 of these pins on a microtiter plate for the execution of library or array syntheses is the most common, as the screening can be simplified with this.

Polymerase chain reaction (PCR) Technique for amplification of DNA, e.g. for obtaining sufficient quantities for identification starting from a few molecules.

Pool sequencing Sequencing of a mixture of peptides by Edman degradation, which permits dominant sequence motifs to be recognized or the quality of smaller libraries to be roughly estimated.

Product matrix Array.

Protecting groups Chemical groups that reversibly block functional groups, in order to prevent them from entering into undesired side reactions.

Resin Synonym for the polymeric support at which the product is synthesized within the framework of solid phase synthesis.

Reverse transcriptase Enzyme that can reverse transcribe RNA, starting from the RNA strand that is required as a template, into DNA.

Ribozyme RNA molecule with catalytic activity.

Screening Testing of chemical compounds with regard to the desired properties (for example biological activity, affinity, conductivity, . . .).

Selectivity Measure of the capability of a compound to only bind to a certain target molecule.

Solid phase synthesis Concept in which the desired product and all of the intermediates remain bound to a polymeric support during the entire progress of the synthesis. All of the reagents are added in a dissolved form, thus excesses and byproducts can be separated through filtration.

Spacer Chemical group that increases the distance between two molecules or molecule parts. Sterically hindered reactions are made easier because of this, i.e. the attack of a voluminous reaction partner is facilitated.

Spot synthesis Solid phase synthesis at certain points (spots) of a two-dimensional surface, for example of a functionalized cellulose membrane.

Sublibrary Subunit of a library, for example after the transformation of a 'mixed' position into a 'defined' position within the framework of the deconvolution.

Tags Identification markers that translate all of the desired information of a library synthesis into a 'language' or code, the analysis of which is possible in a quick and easy way.

Tea bag synthesis Solid phase synthesis technique in which the polymeric support is separated or grouped by sealing it in a plastic mesh.

Toxicity The degree to which a compound is toxic.

INDEX

Note: Figures and Tables are indicated by *italic page numbers*